Materials and Interfaces for Clean Energy

Materials and Interfaces for Clean Energy

Shihe Yang
Yongfu Qiu

JENNY STANFORD
PUBLISHING

Published by

Jenny Stanford Publishing Pte. Ltd.
Level 34, Centennial Tower
3 Temasek Avenue
Singapore 039190

Email: editorial@jennystanford.com
Web: www.jennystanford.com

British Library Cataloguing-in-Publication Data
A catalogue record for this book is available from the British Library.

Materials and Interfaces for Clean Energy

Copyright © 2022 by Jenny Stanford Publishing Pte. Ltd.
All rights reserved. This book, or parts thereof, may not be reproduced in any form or by any means, electronic or mechanical, including photocopying, recording or any information storage and retrieval system now known or to be invented, without written permission from the publisher.

For photocopying of material in this volume, please pay a copying fee through the Copyright Clearance Center, Inc., 222 Rosewood Drive, Danvers, MA 01923, USA. In this case permission to photocopy is not required from the publisher.

ISBN 978-981-4877-66-4 (Hardcover)
ISBN 978-1-003-14223-2 (eBook)

Contents

Preface		xi
1. Introduction		**1**
1.1	Valence Electron Involved Energy Conversion	1
1.2	Enabling Materials and Interfaces	3
	1.2.1 Nanomaterials	3
	1.2.2 Electroactive and Photoactive Materials	4
	1.2.3 Catalytic Materials	5
	1.2.4 Interfaces	5
2. New Material Synthesis		**9**
2.1	Carbon Dots Synthesis	10
	2.1.1 Top-Down Methods	10
	2.1.2 Bottom-Up Methods	11
2.2	Copper Sulfide Nanowire Arrays Synthesis	12
2.3	Silver Sulfide Nanowire Arrays Synthesis	16
2.4	Iron Oxide Nanowire Arrays Synthesis	17
2.5	Copper Hydroxide and Oxide Nanowire Arrays Synthesis	18
2.6	Ultrathin ZnO 1D Nanowire Arrays Synthesis	22
2.7	In Situ Cu_2S/Au Core/Sheath Nanowire Arrays Synthesis	23
2.8	In Situ Cu_2S/Polypyrrole Core/Sheath Nanowire Arrays Synthesis	25
2.9	Ultrathin Bi_2O_3 Nanowire Synthesis	26
2.10	Ultrathin ZnO Tetrapods Synthesis	30
2.11	Ultrathin ZnO Nanotubes Synthesis	32
2.12	Layered Double Hydroxides Synthesis	34
2.13	Binary-Nonmetal TMCs Synthesis	37
2.14	Fully Inorganic Trihalide Perovskite	40
	2.14.1 High-Temperature Hot-Injection Route	41
	2.14.2 Room-Temperature Coprecipitation Method	43
	2.14.2.1 Ligand-mediated reprecipitation method	43

		2.14.2.2	Supersaturated recrystallization process	44
		2.14.3	Droplet-Based Microfluidic Approach	45
		2.14.4	Solvothermal Method	46
	2.15	Conclusion		46
3.	**Interface Engineering for Perovskite Solar Cells**			**49**
	3.1	Introduction of Perovskite Solar Cells		49
	3.2	Importance of Interfaces in Planar p-i-n PSCs		50
		3.2.1	Energy-Level Alignment	51
		3.2.2	Charge Dynamics	52
		3.2.3	Trap Passivation	54
		3.2.4	Ion Migration	54
	3.3	Interface Engineering of Conventional n-i-p Structure PSCs		55
		3.3.1	From Dye-Sensitized Solar Cells to PSCs	55
		3.3.2	Interface between TCO and Oxide Semiconductor ETL	57
		3.3.3	Interface between ETL and Perovskite	59
		3.3.4	Grain Boundaries in the Perovskite Active Layer	64
		3.3.5	Interface between Perovskite and HTL	65
	3.4	Interface Engineering of Inverted p-i-n Structure PSCs		67
		3.4.1	Current–Voltage Hysteresis Problem: From n-i-p to p-i-n Structure PSCs	67
		3.4.2	Interface between TCO and HTL	68
		3.4.3	Interface between HTL and Perovskite	70
		3.4.4	Interface between Perovskite and ETL	75
		3.4.5	Interface between ETL and Metal Electrode	78
	3.5	Interface Engineering of C-PSCs		78
		3.5.1	Brief Introduction of C-PSCs	78
		3.5.2	Interface between Oxide Semiconductor ETL and Perovskite	79
		3.5.3	Interface between Perovskite and Carbon-Based Electrode	82
		3.5.4	New-Type Carbon Electrode Materials	84
	3.6	Conclusion and Perspectives		87

4. Carbon Quantum Dot Luminescent Materials — 91

- 4.1 Introduction — 92
  - 4.1.1 Semiconductor Physics Basics — 92
    - 4.1.1.1 Energy band — 92
    - 4.1.1.2 Intrinsic and extrinsic semiconductors — 93
    - 4.1.1.3 p–n junction — 94
    - 4.1.1.4 Direct and indirect bandgap semiconductors — 95
  - 4.1.2 Semiconductor Luminescence — 97
    - 4.1.2.1 Luminescence process of direct bandgap semiconductor materials — 98
    - 4.1.2.2 Luminescence process of indirect bandgap semiconductor materials — 99
- 4.2 Fluorescence Properties of Carbon Quantum Dots — 99
  - 4.2.1 Introduction of Carbon Quantum Dots — 99
  - 4.2.2 Fluorescence Emission from Bandgap Transitions of Conjugated π-Domains — 101
  - 4.2.3 Fluorescence Emission of Surface Defect-Derived Origin — 106
  - 4.2.4 Up-Conversion Fluorescence — 107
- 4.3 Room Temperature Phosphorescence Properties of CQDs — 108
- 4.4 Thermally Activated Delayed Fluorescence Properties of CQDs — 113
- 4.5 Summary — 114

5. Application in Lithium-Ion Battery — 117

- 5.1 Various Precursors to Hierarchically Porous Micro-/Nanostructures — 118
  - 5.1.1 Metal Hydroxide Precursors — 119
  - 5.1.2 Metal Carbonate Precursors — 121
  - 5.1.3 Metal Carbonate Hydroxide Precursors — 127
  - 5.1.4 Metal–Organic Framework Precursors — 130

		5.1.5	Other Precursors	130
5.2	Application in LIBs			130
	5.2.1	Anode Materials		130
	5.2.2	Cathode Materials		138
5.3	Conclusion			142

6. Application in Perovskite Solar Cells — 145

6.1	Working Principle and Characterization of Solar Cell	146
6.2	Crystal Structures of I-PVKs	148
6.3	Lead-Based Inorganic Perovskites	149
	6.3.1 $CsPbI_3$ Perovskite Solar Cells	149
	6.3.2 $CsPbBr_3$ Perovskite Solar Cells	154
	6.3.3 $CsPbI_{3-x}Br_x$ Perovskite Solar Cells	159
6.4	Lead-Free Inorganic Perovskites	165
	6.4.1 $CsSnI_3$ Perovskite Solar Cells	165
	6.4.2 $CsSnBr_3$ Perovskite Solar Cells	169
	6.4.3 $CsSnI_{3-x}Br_x$ Perovskite Solar Cells	172
	6.4.4 $CsGeI_3$ Perovskite Solar Cells	174
6.5	Perovskite-Derived Materials	175
	6.5.1 Sn-Based Perovskites	175
	6.5.2 Bi-Based Perovskites	179
	6.5.3 Sb-Based Perovskites	181
	6.5.4 Double Perovskites	182
6.6	Issues and Outlooks	185
6.7	Conclusion	186

7. Application in Electrocatalytic Water Splitting — 189

7.1	HER Fundamentals	192
	7.1.1 Hydrogen Adsorption on Catalyst Surface	193
	7.1.2 Reaction Pathways	198
7.2	Binary-Nonmetal TMCs for Hydrogen Evolution	202
	7.2.1 Transition-Metal Sulfoselenides (MSSe)	202
	7.2.2 Transition-Metal Phosphosulfides	209
	7.2.3 Transition-Metal Carbonitrides	215
7.3	Some Basics of OER Catalysis	219

7.4	Transition-Metal-Based Layered Double Hydroxides for Oxygen Evolution		223
	7.4.1	Unary Metal-Based Layered Double Hydroxides	223
		7.4.1.1 VIII group single transition-metal hydroxides/oxyhydroxides	224
		7.4.1.2 V-hydroxides/oxyhydroxides	226
	7.4.2	Binary Metal-Based LDH	226
		7.4.2.1 NiFe LDH	227
		7.4.2.2 Other Ni-based binary metal LDHs	229
		7.4.2.3 Co-based binary metal LDH	231
	7.4.3	Ternary Metal-Based LDH	233
7.5	Mechanistic Studies of OER		235
7.6	Summary and Prospects		239
	7.6.1	Difficulty in Synthesizing Well-Defined Binary-Nonmetal TMC Materials	240
	7.6.2	Difficulty in Characterizing Precise Locations of As-Doped Foreign Nonmetal Atoms in Binary-Nonmetal TMC Materials	240
	7.6.3	Need for Further Improvement in Their HER Electrocatalytic Performance	241

Index   245

Preface

Materials are the bedrock of human society, and major advancements in the field of materials science have often been crowned as a new level in the history of civilization. For instance, the discovery of the periodic table and the understanding of the bonding properties between elements tremendously expedited the efforts to develop novel properties that could be integrated into the compositions and structures of materials. Metal materials, semiconductor materials, ceramic materials, and polymer materials are some of the new materials that are truly a triumph of science and technology of the time. Toward the end of the last century, it was realized that properties of materials can also be engineered by defining their size, morphology, dimensionality, etc.; and this ushered in a new era of materials science in the name of nanomaterials. Carbon pioneered this development path—from 0D fullerenes, 1D bucky tubes to 2D graphenes—sprawling out to myriad materials systems. At present, nanomaterials are arguably the base that integrates nanotechnology, information technology, and biotechnology, the major drivers for the technological development of the day.

At the dawn of the 21st century, energy, which directly impacts the prosperity of modern society, was ranked number one of humanity's top 10 problems for the next 50 years by the late Nobel laureate Richard E. Smalley. Currently, the reliance on fossil fuels raises concerns on the increasing global energy demand, the rapid anthropogenic climate changes, and the growing environmental problems. The grand challenge is to search for viable carbon-neutral sources of renewable energy. By general consensus, solar energy has the largest potential to satisfy the future global need. In fact, as little as 0.01% of the sunlight striking the earth is sufficient to completely satisfy the global power demand in terawatts and power the planet. However, of the total energy produced worldwide, the fraction created from sunlight is currently below 1%, mainly due to the high harvesting cost. So, the real scientific as well as technological question is as to how the abundant terrestrial sunlight can be harvested efficiently and turned into transportable energy in the form of solar fuels at a low cost. This is the exact point when nanomaterials come to our rescue. The rapid development of nanomaterials with unusual properties echoes of Richard Feynman's farsighted lecture: There's Plenty of Room at the Bottom. The ability to engineer nanomaterials' forms allows to achieve the targeted functions of solar fuels in a systemic fashion. Over the past decades, the understanding of form–function relations surrounding nanomaterials has significantly brightened the prospects of the transition of fossil fuels to solar fuels.

This book introduces the latest developments in the application of nanomaterials in the area of clean energy. It discusses the synthesis and interface engineering of new materials as well as carbon quantum dot luminescent materials and describes their applications in lithium-ion batteries, perovskite solar cells, and electrocatalytic water splitting. The book is not intended to be exhaustive and provides only a flavor of the close link between nanomaterials' development and their energy applications from the authors' experience and perspective.

We would like to sincerely thank many colleagues, who and whose work have made this book possible.

Shihe Yang
Yongfu Qiu
June 2021

Chapter 1

Introduction

1.1 Valence Electron Involved Energy Conversion

According to quantum mechanics, valence electrons in molecules and materials can occupy different states with different energies in different nuclear configurations. Thus the change of states of these electrons will involve energy conversion on the scale of chemical bond energy. Such energy conversion underlies essentially all of the modalities of the clean energy devices covered in the subsequent chapters of this book. For the sake of illustration, here we use the following reaction as a simple example:

$$2H_2 \text{ (g, } 10^5 \text{ Pa)} + O_2 \text{ (g, } 10^5 \text{ Pa)} \rightarrow 2H_2O \text{ (l)}$$
$$\Delta H = -571.7 \text{ kJ mol}^{-1} \quad \Delta G = -474.4 \text{ kJ mol}^{-1}$$

The exergonic (also exothermic) reaction involves electron transfer from H_2 to O_2, and if allowed to take place irreversibly, will generally turn chemical energy into heat Q (Fig. 1.1). However, in theory, this reaction can also be carried out reversibly in some way. And in this case, the maximum useful work that can be done is, $\Delta G = -237.2 \text{ kJ mol}^{-1}$.

Materials and Interfaces for Clean Energy
Shihe Yang and Yongfu Qiu
Copyright © 2022 Jenny Stanford Publishing Pte. Ltd.
ISBN 978-981-4877-66-4 (Hardcover), 978-1-003-14223-2 (eBook)
www.jennystanford.com

Irreversible: H_2 (g, 10^5 Pa) + $1/2\ O_2$ (g, 10^5 Pa) $\leftrightarrow H_2O$ (l)
$$Q = \Delta H = -285.9 \text{ kJ mol}^{-1}$$
Reversible: H_2 (g, 10^5 Pa) + $1/2\ O_2$ (g, 10^5 Pa) $\leftrightarrow H_2O$ (l)
$$W_{max} = \Delta G = -237.2 \text{ kJ mol}^{-1}$$

Figure 1.1 Overall reaction.

In fact, here the chemical energy can be converted to useful electric energy by directional movement of the transferred electrons (Fig. 1.2). The maximum extracted electrical energy is $-\Delta G$, which is related to the electromotive force E and the number of transferred electrons n.

Anode: $\quad\quad\quad\quad H_2 \rightarrow 2H^+ + 2e^-$

Cathode: $\quad\quad 0.5O_2 + 2H^+ + 2e^- \rightarrow H_2O$

$$E = \frac{-\Delta G(T, P_0)}{nF}$$

Figure 1.2 Spatially separated half redox reactions (hydrogen oxidation and oxygen reduction to form water).

Photons can also be added as a free energy source to the reaction, for example, to turn the above exothermic reaction in the reverse direction, that is, to split water into hydrogen and oxygen (Fig. 1.3), converting photon energy to chemical energy with a maximum efficiency of η.

Photoanode: $\quad H_2O \xrightarrow{h\nu} 0.5O_2 + 2H^+ + 2e^-$

Cathode: $\quad\quad\quad 2H^+ + 2e^- \rightarrow H_2$

$$\eta = \frac{\Delta G}{2E_{h\nu}}$$

Figure 1.3 Spatially separated half redox reactions (photodriven water splitting).

1.2 Enabling Materials and Interfaces

To realize the valence electron involved energy conversion and extraction described in the last section, suitable devices need to be designed, which in turn require suitable materials and interfaces to transfer, transport, and collect the valence electrons and the opposite charges in the corresponding separate electrodes. In energy conversion processes that involve photons, photoactive materials are also of importance. In the following, we will briefly introduce some important categories of materials and interfaces commonly used in energy conversion.

1.2.1 Nanomaterials

In the past several decades, researchers have been able to design and synthesize nanomaterials with various dimensionalities traversing 0-D nanoparticles, 1-D nanorods/nanowires/nanotubes, 2-D nanosheets, and 3-D complex structures (Fig. 1.4). In retrospect, carbon has played a leading role in those fantastic developments, starting with the discovery of bucky balls (0-D), followed by bucky tubes, and then graphene. These nanomaterials have unique physical and chemical properties associated with their size, shape, and dimensionality. In particular, these nanomaterials have a nanoscale size in at least one of the spatial dimensions reaching the quantum-confinement regime. Quantum mechanics dictates that such nanomaterials will have an energy spectrum displaying characteristic features. For example, the density of states (DOS) of 0-D nanomaterials show line spectra, 1-D nanomaterials show Van't Hoff singularities, while 2-D nanomaterials show step-like features. Such new spectral features provide additional handles to design energy conversion devices, such as quantum dot solar cells, quantum wire radial-type coaxial-junction solar cells, etc. In addition, the nano-sized materials also provide reduced lengths for ion diffusion in batteries, and enlarged surface areas and enhanced activities for promoting surface reaction kinetics in energy conversion catalysis, etc.

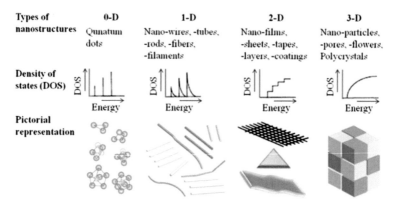

Figure 1.4 Schematic of 0-, 1-, 2-, and 3-dimensional nanostructured materials.

1.2.2 Electroactive and Photoactive Materials

As spelled out from the outset, common energy conversion reactions involve changes of valence electron states, more specifically, changes of oxidation states. In these reactions, electron-rich molecules or materials are more energetic, whereas electron-poor molecules or materials have a lower energy content. Thus the energy conversion involving the transformation between the higher-energy and the lower-energy molecules or materials requires a series of electron transfer steps. Consequently, an efficient energy conversion device should be designed in such a way that the electron transfer steps should be facilitated, which requires not only advanced device design but also superior molecules/materials for fast charge transfer and transport. These species include redox molecules/materials, transport materials, and charge transfer control materials in batteries, solar cells, solar fuel, and light emission devices.

When the energy conversion reactions involve photons, photoactive molecules/materials are also required as essential components in the corresponding devices in addition to more elaborate device designs. First of all, one needs suitable photoabsorber molecules/materials to efficiently harvest photons. Second, the photoabsorber molecules/materials should be able to separate charges using the energy from the absorbed photons, and efficiently transport the separated charges in the opposite directions.

1.2.3 Catalytic Materials

For energy conversion systems involving chemical reactions, oftentimes a sizable reaction energy barrier arises, which when left unaddressed will significantly lower the energy conversion efficiency. For example, although some semiconductors have a suitable energy band position for PEC water oxidation or reduction, the formation of intermediate species on most bare semiconductor surfaces creates a large energy barrier to the generation of oxygen or hydrogen, thus an overpotential is required to drive reaction kinetic rate and then to improve the electrode efficiency. The cocatalyst loading is essential to improve the kinetics of the reaction. The main function of the cocatalyst is to provide redox reaction sites to lower the activation energy of the catalytic reactions. The built-in electric field at the interface between the semiconductor materials and cocatalysts can effectively promote interfacial charge separation. In addition, by efficiently consuming photogenerated carriers, the stability of the photoelectrodes and cocatalyst under light can be improved. It is worth noting that if the deposited cocatalyst has a small size, or a scattered distribution, but has good catalytic activity, then the cocatalyst is optically "transparent" and will not significantly influence the light absorption of semiconductors.

In an ideal water splitting PEC cell, the efficient HER and/or OER cocatalyst must be highly active for their respective reactions, meaning that the cell should be able to produce the hydrogen or oxygen as quickly as possible with a minimal overpotential. In general, metals such as Pt, Rh, Ru, Ir, and Ni as well as transition-metal oxide, sulfide, or phosphide, such as NiO_x, MoS_2, Ni_2P, etc. can function as HER cocatalysts. And the metal phosphates (such as Co-Pi, etc.), metal oxide (such as IrOx, RuOx, CoOx, etc.), Ni-, Fe-, or Co-based nickel (oxy)hydroxide (such as NiOOH, FeOOH, CoOOH, etc.), metal boron (such as Ni–Bi, etc.), layered double hydroxide (LDH), etc. can work as OER cocatalysts.

1.2.4 Interfaces

In energy conversion devices, interfaces play a very important role, where interfacial reactions, charge separation, and charge transfer occur. Considering the role of nanointerfaces in the determination of the local charge separation and transfer, effectively constructing nanointerfaces in the energy conversion system is of great significance for obtaining high-energy conversion efficiency.

Here as an example, we highlight the importance of interface construction by regulating local charge separation and transfer properties. Such interface construction is essential for achieving high performance and stability for hybrid PEC systems. In an ideal PEC cell, there are mainly three types of interfaces: semiconductor–semiconductor, semiconductor–cocatalyst, and photoelectrode–electrolyte (Fig. 1.5). The performance of PEC water splitting depends largely on the nature and capabilities of these interfaces, so it is of great significance to understand the impact of these interfaces on PEC water splitting and further optimize and develop the PEC cells.

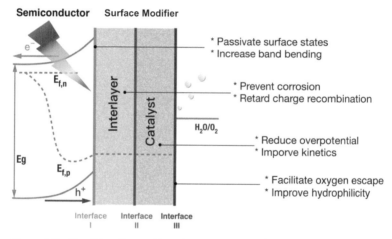

Figure 1.5 Possible effects of interface engineering for PEC water oxidation.

Specially, the interface between the adjacent semiconductor layers is able to provide an additional band bending to promote the charge separation, and to passivate the surface disorder to suppress the charge recombination. The cocatalyst is usually in the form of dispersed nanoparticles or a porous membrane, and serves to effectively extract the charge to avoid the possible accumulated charge induced side reaction on the photoelectrode surfaces. In general, if more efficient cocatalysts are used under optimized conditions, better PEC performance can be obtained. However, the PEC performance of the photoelectrode is very sensitive to the loading amount of the cocatalysts, especially for nanostructured photoelectrodes with a large surface area. This may be because the dense cocatalysts usually lead to a significant increase of the charge transfer, resulting in severe surface charge recombination. In addition, due to the mismatch between the cocatalyst and the semiconductor, there may be many defects and interfacial energy levels at the interface, leading to serious recombination during charge separation and transport processes. In this case, the choice of cocatalysts and the interfacial optimization between cocatalysts and semiconductor materials are extremely important. The photoelectrode-electrolyte interface is another important factor for surface catalysis. A well-managed interface between the photoelectrode and electrolyte can improve the hydrophilicity and facilitate mass transfer, bubble separation and oxygen escape.

The main theme of this book is centered on materials and interfaces for clean energy, wherein nanomaterials play important roles since they have opened new dimensions to manipulate valence electrons more effectively. Following this introductory chapter, six chapters will be dedicated to cover the latest nanomaterials development for applications in the clean energy areas, ranging from new material synthesis and interface engineering, to applications in luminescent materials, lithium-ion batteries, perovskite solar cells, and electrocatalytic water splitting.

Chapter 2

New Material Synthesis

New materials refer to materials with excellent properties that are recently developed or are being developed on the base of traditional materials. The new materials are divided into four categories by composition: metallic materials, inorganic nonmetallic materials (such as ceramics, gallium arsenide semiconductors, etc.), organic polymer materials, and advanced composites. They can also be divided into structural materials and functional materials according to their properties. Structural materials mainly utilize their mechanical and physicochemical properties to meet the performance requirements of high strength, high stiffness, high hardness, high temperature resistance, wear resistance, corrosion resistance, radiation resistance, etc. Functional materials rely on their electrical, magnetic, acoustic, photothermal, and other effects to achieve certain functions, such as composite new materials, superconducting materials, energy materials, intelligent materials, magnetic materials, nanomaterials, etc. Clean energy materials play a particularly important role in energy conversion and environmental protection, and they mainly include luminescent materials, energy storage materials, solar cell materials, hydrogen storage and hydrogen production materials, etc. In this chapter, several representative clean energy materials such as carbon dots (CDs), metal oxides, layered double hydroxides, binary-nonmetal transition-metal compounds (TMCs), and fully inorganic trihalide perovskite nanocrystals are selected for the

Materials and Interfaces for Clean Energy
Shihe Yang and Yongfu Qiu
Copyright © 2022 Jenny Stanford Publishing Pte. Ltd.
ISBN 978-981-4877-66-4 (Hardcover), 978-1-003-14223-2 (eBook)
www.jennystanford.com

introduction and illustration of their synthesis methods according to their morphology and dimensionality (zero, one, two, and three dimensional structures).

2.1 Carbon Dots Synthesis

Since their first discovery in 2006, carbon dots (CDs) as a new family of quantum dots (QDs) have attracted widespread attention and emerged as an excellent fluorescent nanomaterial. Literally, CDs are typically a class of zero-dimensional (0D) nanocarbons, or carbon nanoparticles, which are less than 10 nm across with an obvious crystal lattice parameter of ~0.34 nm corresponding to the (002) interlayer spacing of graphite. CDs can be synthesized by two general strategies: top-down and bottom-up methods. The former involves breaking down or cleaving larger carbon structures to smaller ones via chemical, electrochemical, or physical approaches. The latter is realized by pyrolysis or carbonization of small organic molecules or by stepwise chemical fusion of small aromatic molecules. The precursors and synthesis methods chosen often determine the physicochemical properties of the resulting CDs such as the size, crystallinity, oxygen/nitrogen content, fluorescence characteristics including quantum yield (QY), colloidal stability, and compatibility (Yuan et al., 2016; Yuan et al., 2019).

2.1.1 Top-Down Methods

The first reported fluorescent CDs were synthesized by the top-down method through laser ablation of graphite using argon as a carrier gas in the presence of water vapor, followed by acid oxidative treatment and surface passivation. Thereafter, various top-down synthesis approaches toward CDs were developed by reducing the size of graphite, graphene or graphene oxide (GO) sheets, carbon nanotubes (CNTs), carbon fibers, carbon soot, and other materials that possess a perfect sp^2 carbon structure and lack an efficient bandgap to give fluorescence. For top-down approaches, many methods have been developed to break down the carbon structure into CDs such as arc discharge, laser ablation, nanolithography by reactive ion etching (RIE), electrochemical oxidation, hydrothermal or solvothermal, microwave assisted, sonication assisted, chemical exfoliation, photo-Fenton reaction, and nitric acid/sulfuric acid oxidation. These synthesis methods are invariably complicated and uncontrollable with relative low yields and QYs, some even adding hazards to the environment and thus are not suitable for the large-scale production of CDs with high QY of photoluminescence.

2.1.2 Bottom-Up Methods

On the contrary, the bottom-up methods offer exciting opportunities to controlling the CDs with well-defined molecular weight and size, shape and properties by using elaborately designed precursors and preparation processes. Meanwhile, the bottom-up methods are usually low cost and quite efficient for producing fluorescent CDs on a large scale, a prerequisite for practical applications of these novel CDs. Laser irradiation of toluene, hydrothermal treatment of citric acid or carbohydrates, stepwise solution chemistry methods using benzene derivatives, carbonization of hexa-peri-hexabenzocoronene (HBC), and precursor pyrolysis have been utilized to prepare CDs successfully. Apart from the precursors mentioned earlier, other molecules with abundant hydroxyl, carboxyl, and amine groups are also suitable carbon precursors, such as glycerol, ascorbic acid, and amino acid, which can dehydrate and further carbonize into CDs at elevated temperatures. Yang et al. reported facile and large-scale synthesis of nitrogen-doped CDs with QY of ~80% in aqueous through hydrothermal treatment of citric acid and ethylene diamine, which formed polymer-like CDs first and then carbonized into CDs. Moreover, the emission color and QY of CDs can be effectively tuned by adjusting the ratio of reagents or the amount of ancillary inorganic substrate (e.g., H_3PO_4, KH_2PO_4, NaOH). For instance, Bhunia et al. synthesized highly fluorescent CDs with tunable visible emission from blue to red on gram-scale from different kinds of carbohydrates and dehydrating agent (H_3PO_4, H_2SO_4) at various temperatures. The dehydrating agents and the reaction temperature were varied to control the nucleation and growth kinetics. This synthetic method could restrict the particle size to smaller than 10 nm and afford an appropriate chemical composition of CDs for tunable emission with a relative high fluorescence QY.

2.2 Copper Sulfide Nanowire Arrays Synthesis

The last two decades have witnessed an explosive development in the bottom-up synthesis of nanostructured materials, ranging from fullerenes and CNTs to nanoparticles and nanowires. Harnessing these novel nanomaterials for device applications, however, necessitates their assembly into ordered nanostructured arrays or higher level nanoarchitectures. The assembled nanostructures are expected to not only have an enhanced collective response to external stimuli, for example, electromagnetic fields or chemical species, but also build synergetic multifunctionalities into an integral system of a device.

The anisotropy of nanowires renders them more difficult to assemble than spherical nanoparticles into a large structure of any dimensionality, be it a horizontal or a vertical array. Whereas post-assembly strategy has been used to obtain horizontal nanowire arrays via solvent flow through microfluidic channels and by the Langmuir–Blodgett deposition, it is difficult to apply to vertical nanowire arrays. Yet vertical nanowire arrays display many desirable characteristics for device integrations, such as enhanced capacity and rate capability in lithium-ion batteries and improved light coupling and electron lifetime in dye-sensitized solar cells.

Traditionally, vertical nanowire arrays were fabricated by the use of templates such as membranes of anodic aluminum oxide (AAO) and track-etched polycarbonate. Ozin et al. and Martin et al. first used nanoporous membranes as templates to prepare arrays of nanotubes and nanorods. Although the template method is general and flexible, the removal of template turned out to be a hassle and often accompanied by aggregation. Vertical nanowire arrays can also be fabricated by vapor–liquid–solid (VLS) and oxide-assisted processes. However, the VLS approach and its analogues generally require catalysts, high temperature, or both, which may limit some of their applications.

For the past few years, researchers have been working on an alternative strategy to integrate the growth and assembly of nanowires on metal substrates. The defining characteristic of our strategy is that the metal substrate itself is part of the reactants to sustain the nanowire growth into a vertical array without using templates and catalysts. The parallel synthesis and assembly of nanowire arrays obviate the need for post-assembly and have the advantages of simplicity, mild reaction conditions, and low cost, opening up a promising avenue for engineering nanodevices. In this section, copper sulfide nanowire arrays growing on a copper substrate by the gas–solid (GS) approach will be discussed as an example (Zhang and Yang, 2009).

The reaction conditions were unusually simple and mild: by exposing a well-cleaned copper foil into an atmosphere of O_2 and H_2S at room temperature, we obtained a black film homogeneously covering the copper substrate, which turned out to be a uniform array of Cu_2S nanowires. The array consists of straight Cu_2S nanowires (Fig. 2.1A) with a diameter of ~50–70 nm and length of ~7–16 µm, which are aligned approximately perpendicular to the substrate surface. These nanowires are crystalline, and some of them have a thin layer of oxide coating (fcc Cu_2O) at the surface (Fig. 2.1B) due to the presence of reactive O_2.

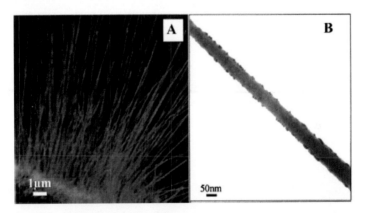

Figure 2.1 (A) SEM image of a Cu_2S nanowire array grown on a copper foil. (B) TEM image of a single Cu_2S nanowire.

In order to understand the intriguing nanowire growth mechanism, we examined early stages of the Cu_2S nanowire development. When the reaction time between the copper surface and the gas mixture of H_2S/O_2 was 10 min, mainly a Cu_2O layer was observed. Cu_2S nuclei started to appear on the Cu_2O layer for reaction time >20 min. After nucleation, the Cu_2S nanostructures then evolved into nanowires and grew mainly along the direction perpendicular to the copper surface (~100 min). A more extended growth period resulted in well-aligned Cu$_2$S nanowires as shown in Fig. 2.2A.

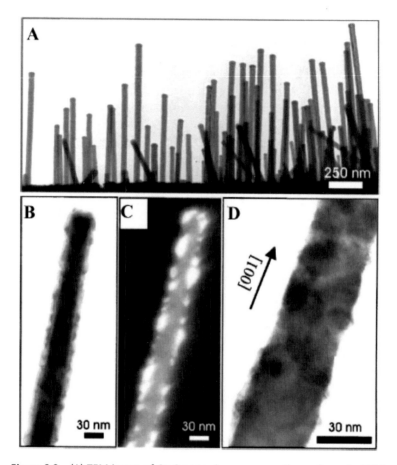

Figure 2.2 (A) TEM image of Cu$_2$S nanowires grown on the surface of a TEM copper grid. (B–D) Core/shell structure, dark-field image using (200) Cu$_2$S diffraction (bright dot contrast comes from the Cu$_2$O shell), and HRTEM image of a Cu$_2$S nanowire.

The nanowires are not only straight but also uniform in diameter (d = ~40 nm) and have lengths of a few hundred nanometers. It seems that there is an induction period for the nucleation of incipient Cu$_2$S nanowires, after which the 1D growth is much faster. Interestingly, each nanowire consists of an inner single-crystalline core and an outer shell as can be seen more clearly in Fig. 2.2B. By combination of the selected area electron diffraction (SAED) and dark-field imaging, it was verified that the inner core of single-crystalline Cu_2S is enclosed by an outer shell of polycrystalline Cu_2O (Fig. 2.2C,D).

It was found that O_2 is essential to the formation of the Cu_2S nanowires on a copper surface despite the reaction $2Cu(s) + H_2S(g) = Cu_2S(s) + H_2(g)$ being exergonic at room temperature ($\Delta G° = -52.8$ KJ/mol). The nanowire formation is believed to start with an oxidation reaction $2Cu(s) + 1/2O_2(g) = Cu_2O(s)$, followed by sulfidation $Cu_2O(s) + H_2S(g) = Cu_2S(s) + H_2O(g)$. These processes probably correspond to the induction period for the nanowire nucleation mentioned earlier. The well-established VLS growth mechanism is not supported by our experiments; we never observed a drop-like feature at the nanowire tips. Nevertheless, the nanowire assembly site has to be at one of the two ends, the root or the tip. The root-growth mechanism might be at work because the close-by copper surface can provide the necessary feeding materials for the nanowire growth. In this mechanism, however, the assembly process would be kinetically hindered at this site. More important, the root-growth mechanism cannot explain why the Cu_2S nanowires thicken or become thin as the reaction proceeds. On the other hand, the tip-growth mechanism appears to explain all the facts observed so far. Here the growth is on the nanowire tip, and the feeding stocks are from the roots. This demands a reasonably efficient transport channel for the feeding stocks, e.g., copper ions, across or through the nanowires. The nanowire growth mechanism is summarized in Fig. 2.3. After the nucleation of Cu_2S from sulfidation of Cu_2O (panels A–C), Cu_2S proceeds with 1D growth vertically (panels C, D) driven by its crystalline anisotropy. To sustain the nanowire growth, O_2 is adsorbed on the nanowire tip and reduced, leaving electron holes and Cu^+ vacancies on the nanowire tip. Both the electron holes and the Cu^+ vacancies migrate through the Cu_2S nanowire from the tip down to the root and are annihilated there. The Cu_2O species thus formed on the nanowire tip is then converted to Cu_2S by the reaction with H_2S. The nanowire assembly continues in this way on the tip by interweaving the sulfur and copper layers in alternation. A similar mechanism was proposed for the growth of Cu_2S thin films on copper surfaces.

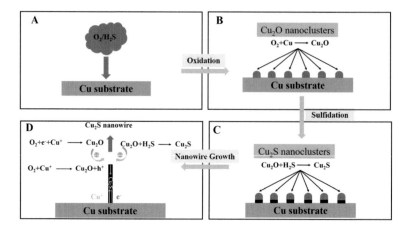

Figure 2.3 Schematic tip-growth mechanism of Cu$_2$S nanowire arrays on copper surface: (A) Cu surface in the presence of O$_2$ and H$_2$S; (B) formation of Cu$_2$O nanoclusters; (C) formation of Cu$_2$S nanoclusters; (D) tip growth of Cu$_2$S nanowires.

2.3 Silver Sulfide Nanowire Arrays Synthesis

The GS approach was also successfully used to prepare Ag$_2$S nanowires on silver substrates. When a pre-oxidized silver substrate was exposed to an atmosphere of an O$_2$/H$_2$S mixture at room temperature or slightly above, Ag$_2$S nanowires were abundantly produced with diameters of 40–150 nm and aspect ratios up to ~1000. Our results suggest that the diameter and morphology of the nanowires are mainly controlled by the nanowire growth rate and the diffusion rate of Ag$^+$ ions from the substrate to the nanowire active site. At low temperatures, Ag$^+$ diffusion in the growing nanowire is expected to be slow. As the nanowire grew longer and longer, the Ag$^+$ supply became shorter and shorter to sustain the uniform nanowire growth, which results in the formation of a nanocone structure with a sharp tip. As the temperature was increased, however, the atomic diffusion rate of Ag increased, which provided a sufficient amount of Ag$^+$ at the nanowire tip for the growth. Consequently, the individual nanowire diameters were more uniform with relatively blunt tips. This strongly supports the tip-growth mechanism of the Ag$_2$S nanowires.

2.4 Iron Oxide Nanowire Arrays Synthesis

One of the problems with the GS growth of metal oxide nanowire arrays is that metal ion diffusion is commonly slower in the metal oxides than in the corresponding sulfides. Increasing the reaction temperature, use of appropriate oxidative gases, and judicious choice of metal/metal oxide systems could be viable solutions. Indeed, ZnO nanobelt arrays and α-Fe$_2$O$_3$ nanobelt/nanowire arrays have been successfully synthesized from and on zinc and iron substrates, respectively, by the GS reaction approach at elevated temperatures but still lower than or close to the melting points of the metals.

Direct thermal oxidation under a flow of O$_2$ produced α-Fe$_2$O$_3$ nanobelts in the low-temperature region (~700°C), but cylindrical nanowires were formed at relatively higher temperatures (~800°C). Both nanobelts and nanowires are mostly bicrystallites with a length of tens of micrometers, which grow uniquely along the [110] direction. The growth habits of the nanobelts and nanowires in the two temperature regions indicate the role of growth rate anisotropy and surface energy in dictating the ultimate nanomorphologies.

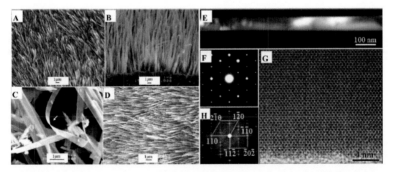

Figure 2.4 α-Fe$_2$O$_3$ nanobelt and nanowire arrays on iron substrates: (A,B) Top view and side view SEM images of a nanobelt array grown at 700°C; (C) high-magnification SEM image of the nanobelts transferred from the substrate; (D) bird's eye view SEM image of a nanowire array synthesized at 800°C; (E–H) dark-field TEM image, SAED pattern, and HRTEM (FFT) image of an α-Fe$_2$O$_3$ nanowire.

Figure 2.4A,B shows SEM images for the samples prepared at 700°C. Clearly, only wire-like features were produced, which are aligned in a dense array approximately perpendicular to the substrate surface with a uniform coverage. Close examination of the products rubbed from the substrate surface revealed that the wire-like features are actually nanobelts with a thickness of several nanometers (Fig. 2.4C). In general, the nanobelts obtained under our growth conditions are about 5–10 nm in thickness, 30–300 nm in width, and 5–50 μm in length.

Shown in Fig. 2.4D–H are nanowires prepared at 800°C. The diffraction contrast (E) suggests a cylindrical wire morphology, while the SAED pattern (F) indicates a bicrystal structure, confirmed by the HRTEM image (G) and FFT (H). It is clear that most of the synthesized α-Fe$_2$O$_3$ nanobelts and nanowires have a bicrystal structure, and the nanostructures grow uniquely along the [110] direction.

The peculiar morphologies from the 1D growth of α-Fe$_2$O$_3$ lend further support to the tip-growth mechanism. First, the 1D growth of α-Fe$_2$O$_3$ is along the [110] direction. We suggested that the (110) plane, being O-rich and Fe-deficient, is more reactive, driving more efficient Fe^{3+} transport and thus the belt growth in the [110] direction. Second, narrowing and thinning of the nanobelts occur with the lapse of reaction time due to a short supply of Fe^{3+} ions for long nanobelts. Finally, we found that low-temperature synthesis prefers the formation of nanobelts, and high-temperature synthesis favors the growth of nanowires. This nanobelt-to-nanowire morphology transition, although related to surface energy, is also at least partly caused by different ion diffusion rates along different crystal directions at different temperatures; in particular, accelerated Fe^{3+} diffusion along the wire direction contributes to the formation of nanowires in lieu of the tapered nanobelts.

2.5 Copper Hydroxide and Oxide Nanowire Arrays Synthesis

The nature of solution–solid (SS) reaction assembly of inorganic nanowire arrays suggests different mechanisms at work from those of the GS method because here the solvent can be an effective carrier for reactive ion transport. The SS method consists of an initial nucleation at the substrate and the subsequent 1D vertical growth with the supply of metal ions generated from the metal substrate and transported to the nanowire tip.

By using the SS approach with either Cu$_2$S nanowires as sacrificial precursor or bare copper surface under O$_2$, we obtained well-aligned Cu(OH)$_2$ nanoribbon arrays on copper foil in an ammonia alkaline solution. For the bare copper surface under O$_2$, the reactions probably proceed as follows:

$$Cu + O_2 + NH_3 \rightarrow Cu[NH_3]_n^{2+} \quad (2.1)$$

$$[Cu(NH_3)_n]^{2+} \rightarrow [Cu(NH_3)_{n-1}(OH)]^+ \rightarrow [Cu(NH_3)_{n-2}(OH)_2] \rightarrow \rightarrow ...$$

$$[Cu(OH)_n]^{(n-2)-} \rightarrow Cu(OH)_2 \text{ (nanoribbon)} \quad (2.2)$$

Figure 2.5 shows a TEM image of the Cu(OH)$_2$ nanoribbons grown directly from the copper grid at room temperature for 12 h. The thinning in the bend and wring sections, which would not occur for cylindrical nanowires (see arrows in A), demonstrates the ribbon-like morphology. These nanoribbons have widths of 20–130 nm and thicknesses of a few nanometers along the whole lengths of several tens of micrometers. Figure 2.5B shows a low-magnification TEM image of a typical single Cu(OH)$_2$ nanoribbon with a width of 60 nm. Although some tiny cavities can be seen on the surface, the nanoribbon is actually a single crystal, as is evidenced by the corresponding ED pattern (inset of panel B). The spotted selected area ED pattern appears to be associated with the [010] zone axis of orthorhombic Cu(OH)$_2$, and it shows that the growth direction of a single nanoribbon is [100]. An HRTEM image of the single nanoribbon is shown in Fig. 2.5C. The clear fringes with a spacing of 0.28 nm match well the distance between the (110) crystal planes. As the inset of Fig. 2.5C shows, the [110] direction makes an angle (~20°) with the axis of the nanoribbon, which is close to the angle between [100] and [110] of orthorhombic Cu(OH)$_2$. This further confirms the nanoribbon growth direction of [100].

The formation of the Cu(OH)$_2$ nanoribbons can be explained by a type of coordination assembly. The idea is schematically shown in Fig. 2.5D. As is well known, Cu^{2+} prefers square-planar coordination by OH$^-$ (a), and this leads to an extended chain (b), which is actually the final nanoribbon direction. The chains can be juxtaposed together through coordination of the OH$^-$ group to the d$_{z^2}$ orbital of

Cu^{2+}, forming a 2D structure (c), which controls the width of the final nanoribbon. Finally, the 2D layers are stacked together through the relatively weak hydrogen bonds and become a 3D crystal. Essentially, the bonding nature in the different crystal directions dictates the growth rates and thus the final nanoribbon morphology.

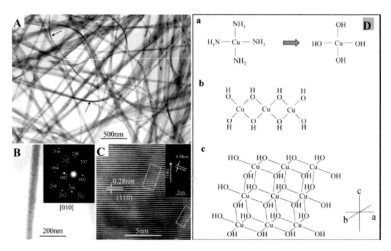

Figure 2.5 (A) TEM image of Cu(OH)$_2$ nanoribbons grown on a copper grid. (B) TEM image and ED pattern (inset) of a single Cu(OH)$_2$ nanoribbon. (C) HRTEM images of a nanoribbon and its boundary section (inset). (D) Schematic illustration of the coordination assembly of the Cu(OH)$_2$ nanoribbons.

Through further heat treatment at 120–180°C with a constant flow of N$_2$, Cu(OH)$_2$ nanoribbon arrays were transformed to CuO without obvious morphological changes.

$$Cu(OH)_2 \xrightarrow{\Delta} CuO + H_2O \qquad (2.3)$$

Shown in Fig. 2.6 are SEM and TEM images of Cu(OH)$_2$ and CuO nanoribbon films. The Cu(OH)$_2$ nanoribbons cover the copper surface uniformly, smoothly, and compactly (A) and are roughly aligned perpendicular to the copper surface (B). After heat treatment, CuO nanoribbon arrays were formed. As shown in Fig. 2.6C, although somewhat shrunken, the ribbon-like morphology is well preserved. The SAED pattern together with the X-ray diffraction (XRD) profile confirms the complete conversion of the Cu(OH)$_2$ nanoribbons to CuO nanoribbons. Shown in Fig. 2.6E is the SEM image of a CuO nanoribbon array obtained after reaction at 5°C for 96 h. Apparently,

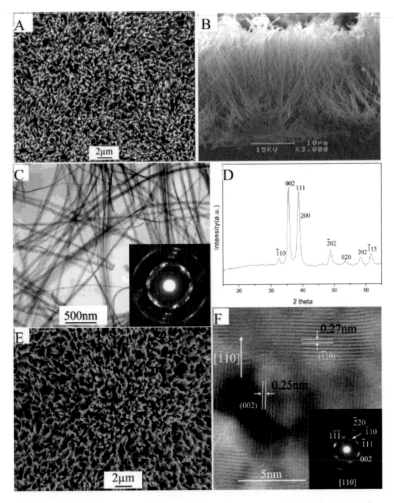

Figure 2.6 (A) Top view and (B) side view SEM images of Cu(OH)$_2$ nanoribbon arrays. (C) TEM image and (D) XRD pattern of CuO nanoribbon arrays transformed from Cu(OH)$_2$. (E) Top view SEM image of the CuO nanoribbon array. (F) HRTEM image and SAED pattern of a single CuO nanoribbon.

the CuO nanoribbons still keep the aligned structure on the Cu substrate. The HRTEM image in Fig. 2.6F shows that the single CuO nanoribbon is a crystal though the crystallinity is not perfect with wavy or discontinuous fringes. With stronger oxidants and bases, such as (NH$_4$)$_2$S$_2$O$_8$ and NaOH, a series of 1D nanostructured arrays of Cu(OH)$_2$ and CuO were obtained. In general, the formation of Cu(OH)$_2$

and CuO nanostructures on copper surface involves complicated reactions of polycondensation and dehydration through competing processes of dissolution–precipitation. The morphological evolution is highly dependent on the synthetic conditions. Our use of a strong oxidant and base appears to make possible the formation of some intermediate phases such as $Cu(OH)_2$ nanofibers and nanotubes, and CuO nanosheets, nanostrips, and nanobelts.

2.6 Ultrathin ZnO 1D Nanowire Arrays Synthesis

Hydrothermal treatment of a zinc substrate in an ammonia/alcohol/water mixed solution resulted in ultrathin ZnO nanofibers aligned on the zinc substrate (Zhang and Yang, 2009; Zhang et al., 2014). The ZnO nanofibers are ultrathin (3–10 nm) with a length of ~500 nm. The growth mechanism of the ultrathin ZnO nanofibers appears to be very similar to that of $Cu(OH)_2$ nanobelts described earlier. In developing the SS approach, we sought to get more handles on the control over nanowire growth on the substrate by applying electrical potentials. With the electrochemical deposition method, both ultrathin ZnO nanobelts and ZnO nanorods as thin as 8 nm in ordered arrays on the Zn cathode were synthesized in alkaline solutions of amine-alcohol mixtures. Although ZnO was found to be deposited on both anode and cathode of Zn foils, vertically arrayed ZnO nanorods were only produced on the cathode. Our results showed that the cathode can nucleate ZnO through oxidation of Zn by electrolysis-generated O_2. The growth of ZnO nanorods is sustained either by electrochemical oxidation of the anode or by chemical oxidation of the cathode. Figure 2.7 shows the effect of applied electric potentials. No 1D nanostructure of ZnO can be found in the absence of the applied electric potential (A). When an electric potential of 0.4 V was applied, vertically arrayed ZnO nanorods were produced on the cathode (B). Further increase in the potential to 2.5 V resulted in the formation of the slightly longer ZnO nanobelts (200 nm), which were assembled roughly into a bundle-like morphology (C). When the potential was finally elevated to 3.8 V, ZnO nanobelts as long as 1 µm were obtained, and they were assembled completely into dense nanobundles to minimize surface energy (D). The 1D arrays of ZnO are thought to form by the following cathode reactions in conjunction with in situ nucleation and the subsequent 1D tip growth:

$$Zn + O_2 + 2H_2O + 2e^- \rightarrow Zn(OH)_4^{2-} \quad \varphi^o = 0.554 \text{ V} \quad (2.4)$$

$$Zn(OH)_4^{2-} \rightarrow ZnO + H_2O + 2OH^- \quad (2.5)$$

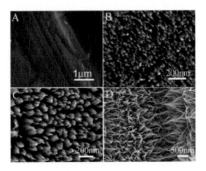

Figure 2.7 SEM images of ZnO products on a Zn cathode at various applied electric potentials: (A) 0; (B) 0.4; (C) 1.5; (D) 3.8 V. Other conditions: [H_2O_2] 22 mM; pH 12.

2.7 In Situ Cu$_2$S/Au Core/Sheath Nanowire Arrays Synthesis

Surface modification of nanowire arrays with a conducting nanolayer is expected to add more functionalities and may even lead to completely new nanocomposite materials, which may be useful in sensors, detectors, rechargeable batteries, and energy and information processing devices.

The gold coating on the Cu$_2$S nanowires was accomplished via the redox reactions of galvanic cells. During the redox processes, the Cu$_2$S nanowires act as cathodes where Au deposition occurs, and the copper substrate serves as an anode where copper dissolution takes place. The electrolyte was the coating solution containing HAuCl$_4$. At least during the initial phase, the Cu$_2$S nanowires as a fairly good electric conductor short the circuits, ensuring the continuation of the gold-coating reactions. Later on, the gold sheath itself should constitute a much better conducting path. Because the reduction potential of AuCl$_4^-$/Au is much higher than that of Cu^{2+}/Cu, the deposition of Au on the Cu$_2$S nanowires and the simultaneous dissolution of Cu from the copper substrate were spontaneous. Typical TEM images of the Cu$_2$S nanowires are shown in Fig. 2.8 before (A) and after the gold coating (B, C). Before coating, the nanowires are ~60 nm in diameter and have a relatively smooth surface (A). On immersion of the nanowires in a coating solution for 1 h, the diameter of the nanowire was increased to ~100 nm (B, C). Although the nanowire surface became rougher after coating, it appears to be compactly covered by the gold deposit along the entire length of the nanowire. The roughness of the nanowire surface indicates that the gold coating is polycrystalline probably because Au was deposited on the Cu$_2$S nanowire surfaces in the form of Au nanocrystals aggregated to form a continuous coating layer. The ED pattern in Fig. 2.8D confirms the polycrystalline nature of the outer sheath of Au (ring pattern) as well as the single-crystal nature of the inner core of Cu$_2$S (discrete diffraction spots).

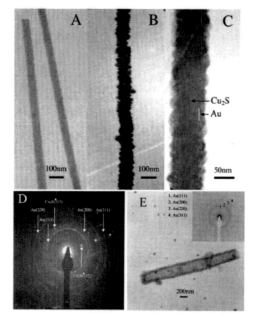

Figure 2.8 TEM images of Cu$_2$S nanowires before (A) and after (B, C) Au coating. (D) ED pattern of a single Cu$_2$S/Au core/sheath nanowire. (E) TEM image and ED pattern of a gold nanotube after removal of the Cu$_2$S core from a Cu$_2$S/Au core/sheath nanowire.

For the Cu_2S/Au core/sheath nanowires, the core of Cu_2S could be removed by etching with acid to form gold nanotubes. Figure 2.8E shows the TEM picture of such a gold nanotube obtained after immersion in an acidic solution for an extended period of time. It has a diameter of ~300 nm and a wall thickness of 50 nm. Clearly, one end of the gold tube is closed, and the other is open. The ED pattern (inset) shows the polycrystalline fcc structure of the gold tube and the absence of diffraction features from Cu_2S originally residing in the core.

2.8 In Situ Cu_2S/Polypyrrole Core/Sheath Nanowire Arrays Synthesis

Polypyrrole (PPY) as a typical conducting polymer (CP) offers reasonably high conductivity and good environmental stability. However, the inherent intractability of PPY has prevented it from being blended with other materials to form nanocomposites using conventional techniques. Although nanoparticles of inorganic core and CP shell were synthesized by coating and catalytic in situ polymerization, the CP often precipitated and could not be easily separated from the core/shell nanoparticles. By using an interfacial polymerization technique, we succeeded in coating Cu_2S nanowires with a homogeneous and well-adhered layer of PPY 20–50 nm thick. The PPY coating could be controlled at the interfacial layer of chloroform and water by polymerization time, pyrrole concentration, and pyrrole-to-oxidant ratio.

The PPY-coating result is shown in Fig. 2.9. The arrayed Cu_2S nanowires used for PPY coating are ~5 μm long, isolated from each other, and roughly perpendicular to the substrate surface (A). After in situ pyrrole polymerization for 2 h, the nanowire morphology is clearly preserved (B). In addition, no PPY precipitate can be seen in the gaps between the PPY-coated Cu_2S nanowires, indicating that the thin PPY layers are closely attached to the Cu_2S cores. The TEM image in Fig. 2.9C displays a core/shell nanowire after pyrrole polymerization. One can see that the Cu_2S nanowire core with a diameter of 65 nm is closely encapsulated by a conformal coating layer of PPY 20 nm thick. The SAED in the inset of Fig. 2.9C proves that the single-crystalline structure of the inner core of Cu_2S remained unchanged. We believe that the PPY layers will improve not only the conductivity but also the stability and mechanical properties of the Cu$_2$S nanowires, which are otherwise environmentally sensitive and mechanically brittle.

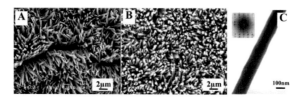

Figure 2.9 SEM images of a bare Cu$_2$S nanowire array (A) and a PPY-coated Cu$_2$S nanowire array (B) obtained with a pyrrole polymerization time of 2.0 h. (C) TEM image and SAED pattern a single PPY-coated Cu$_2$S nanowire.

2.9 Ultrathin Bi$_2$O$_3$ Nanowire Synthesis

Nanowires of Bi$_2$O$_3$ have been an interesting research topic recently, because Bi$_2$O$_3$ is a material with important properties, such as semiconductor characteristics (E_g is 2.85 eV for α-Bi$_2$O$_3$ and 2.58 eV for β-Bi$_2$O$_3$), high refractive index ($n_{\delta-Bi_2O_3}$ = 2.9) and high dielectric permittivity (ε_r = 190), high oxygen conductivity (1.0 S/cm) as well as marked photoconductivity and bright photoluminescence. Although some reports on the synthesis of Bi$_2$O$_3$ nanowires have appeared, the control over the direct synthesis of α- and β-Bi$_2$O$_3$ nanowires has not been reported. In this context, we herein report our successful control over the phase-selective synthesis of α- and β-Bi$_2$O$_3$ nanowires by an oxidative metal vapor transport deposition technique, which is difficult to achieve by other methods. We further propose the growth mechanisms on the basis of a series of careful studies on the nanowire growth conditions (Qiu et al., 2006; Qiu et al., 2011).

The setup for nanowire growth is shown in Fig. 2.10, which consists of a horizontal tube furnace with 120 cm in length and 10 cm in diameter, a quartz tube (100 cm in length and 5 cm in diameter) with two gas inlets and one outlet, and a gas flow control system. A typical synthesis of bismuth oxide nanowires was performed as follows. To begin, 1.5 g bismuth powder using as metal source was loaded onto a quartz substrate and put at the high-temperature zone. A piece of Al foil serving as the product deposition substrate was positioned at the open end of the quartz tube. The quartz tube was then mounted in the middle of the tube furnace. For removing any air in the system, the quartz tube was purged for 40 min by a flow of high-purity nitrogen (>99.995%, 600 sccm). Afterward, the N_2 flow was kept at 600 sccm, and at the same time, the tube furnace was heated to 800°C at the rate of 30°C/min. The oxygen was injected into the quartz tube by pulse or continuous injection mode at the 300°C or higher temperature zone. After being held at this temperature for 8 h, the system was allowed to cool down naturally to room temperature in a nitrogen flow with 100 sccm. Light yellowish products deposited on the substrate were collected carefully for characterizations and measurements.

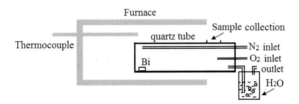

Figure 2.10 Schematic diagram of the setup for the Bi_2O_3 nanowires synthesis.

The nanowire products deposited on the Al foil substrate at two different temperature zones of (i) 450–550°C and (ii) 250–350°C are denoted as Bi_2O_3-NW500 and Bi_2O_3-NW300, respectively. The XRD patterns demonstrate that Bi_2O_3-NW500 takes the α-Bi_2O_3 crystal structure, whereas Bi_2O_3-NW300 has the metastable β-Bi_2O_3 crystal structure. Scanning electron microscope (SEM) images in Fig. 2.11 show that the Bi_2O_3-NW500 consists of very uniform nanowires, 80 to 200 nm in diameter and several hundreds of micrometers in length. On the other hand, the Bi_2O_3-NW300 contains exclusively ultrathin nanowires ∼7 nm in diameter and several μm in length. Therefore, the nanowire products with the two crystal structures could be selectively obtained under our present experimental conditions. The energy dispersive X-ray spectroscopy (EDS) profiles in Fig. 2.12 show that a little droplet on the tip of nanowire is found. The droplet is composed of 100% Bi metal; however, the body of the nanowire is composed of Bi and O with the ratio of Bi to O approximately equal to 2 : 3.

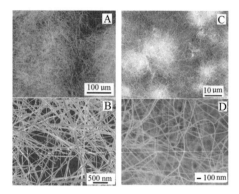

Figure 2.11 SEM images of (A) the as-prepared Bi$_2$O$_3$-NW500 (α-Bi$_2$O$_3$ nanowires) and (B) corresponding enlarged image; (C) the as-prepared Bi$_2$O$_3$-NW300 (β-Bi$_2$O$_3$ nanowires) and (D) the corresponding enlarged image.

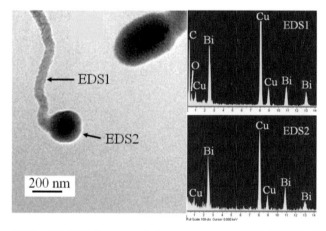

Figure 2.12 (A) TEM images of the as-prepared Bi$_2$O$_3$-NW300 (β-Bi$_2$O$_3$ nanowires). (EDS1, EDS2) Energy dispersive X-ray spectroscopy (EDS) profiles of a nanowire shown in (A) in two areas marked by arrows.

After the experimental conditions such as nitrogen flow rate, sample collection zone, oxygen feeding site, evaporation temperature, and reaction time have been systematically studied, the mechanisms for the growth of α- and β-Bi$_2$O$_3$ nanowires were proposed.

The growth of the β-Bi$_2$O$_3$ nanowires is believed to follow the route of oxidative vapor–transport–oxidation–deposition in a similar vein to the VLS mechanism but with a difference of a coupled oxidation reaction, as we reported previously. Briefly, the Bi vapor is first generated at high temperature perhaps with β-Bi$_2$O$_3$ seeds attached to Bi droplets (800°C or above), followed by the rapid transfer to low-temperature zones by a flow of nitrogen, and then further oxidized by the injected oxygen, and sustains the continued growth of β-Bi$_2$O$_3$ nanowires. Thus, it is not surprising that sizes and morphologies of the β-Bi$_2$O$_3$ nanowires can be controlled by varying the evaporation temperature, nitrogen flow rate, oxygen injected mode and concentration, and collection zone as we found experimentally.

Figure 2.13 is a schematic diagram for the vapor–solid (VS) growth mechanism of the α-Bi$_2$O$_3$ nanowires. First, the Bi vapor is thought to react with the oxygen to form BiO$_x$, which then nucleates at 450~550°C and is further oxidized to α-Bi$_2$O$_3$. The α-Bi$_2$O$_3$ nanowires are then grown in this way (VS) with the fastest growing direction along [100]. When the nanowire diameter is small, the cross section is circular, and the diameter is uniform across the entire nanowire so as to minimize surface energy. When the nanowire diameter is increased to more than about 1 μm, the nanowires become faceted and the cross section is changed to a square shape because the bulk energy becomes more important than the surface energy. In addition, the nanowire tips become tapered probably to minimize the high energy surface.

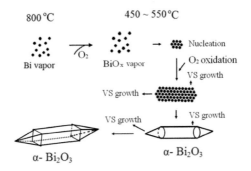

Figure 2.13 Schematic diagram showing the growth mechanism of α-Bi$_2$O$_3$ nanowires. VS means vapor–solid mechanism.

2.10 Ultrathin ZnO Tetrapods Synthesis

Previously, ZnO tetrapods were normally prepared by thermal evaporation method, in which evaporation of Zn metal, oxidation, nucleation, and growth all took place at high temperatures (>600°C). We argue that the tetrapods, once nucleated, would grow rapidly in the high-temperature region and turn to exclusively large sizes as often obtained. Now suppose that the growth can somehow be quenched in a controlled fashion, and then one should be able to tailor the size and morphology of the tetrapods. In the present chapter, we have developed a fast-flow technique for the controlled growth of ZnO nano-tetrapods. This technique combines metal vapor transport, oxidative nucleation/growth, fast-flow quenching, and water-assisted cleaning. Through a series of systematic studies, a number of key factors have been identified that sensitively affects the growth of the ZnO nano-tetrapods. The spatial–temporal separation of ZnO tetrapods formation at high temperatures and quenching processes at low temperatures enabled by the fast-flow technique allows the ZnO tetrapods to be effectively formed and their sizes varied under control and further, collected at relatively low temperatures. In particular, the ZnO nano-tetrapods with arm diameters down to 17 nm and a length of several hundred nanometers have been successfully prepared for the first time using this technique (shown in Fig. 2.14) (Qiu and Yang, 2007).

The process chart for the tetrapod growth can be envisaged in Fig. 2.15. First, Zn is evaporated at high temperatures (≥700°C) to give sufficiently high Zn vapor pressure. This Zn vapor is then transported to the lower temperature zones by a flow of nitrogen and formation of Zn clusters ensues. At a suitable temperature zone (≥500°C), a flow of oxygen is injected, setting off the nucleation and growth of ZnO tetrapods. The size of the tetrapods is dependent on the local temperature and Zn vapor pressure. High temperature and Zn vapor pressure lead to rapid growth into large tetrapods, whereas the opposite may result in too slow a nucleation/growth rate. A compromise is to select a suitable O_2 injection temperature not too high and not too low. More important, the tetrapods formed at higher temperature zones are rapidly transported to lower temperature zone by the flowing nitrogen stream, thus effectively slowing down and terminating their growth. One problem encountered with this growth scheme is the aggregation mentioned above, but it has been solved nicely by adding water vapor at a suitable temperature zone, which represses aggregation and purifies the ZnO tetrapods by adsorption on the growing particles.

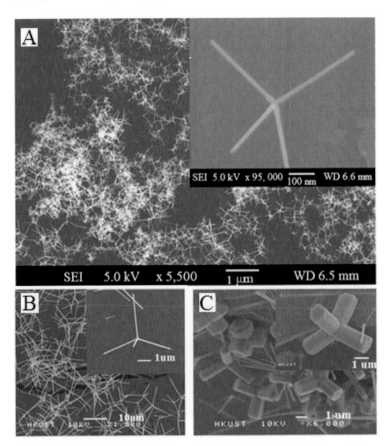

Figure 2.14 SEM images of ZnO tetrapods collected in temperature zones of 100°C (A), 300°C (B), and 500°C (C). Insets in (A), (B), and (C) are the corresponding enlarged images.

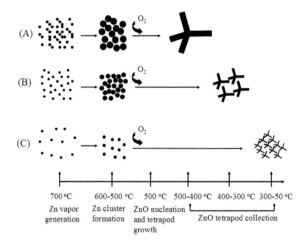

Figure 2.15 Schematic illustrating possible formation processes of ZnO tetrapods. The routes (A), (B), and (C) are associated with high, medium, and low Zn vapor pressures, respectively.

2.11 Ultrathin ZnO Nanotubes Synthesis

The Kirkendall effect, originally observed in bulk diffusion couples of Cu and CuZn, is nowadays employed to fabricate hollow spherical nanocrystals and their chains. Up till now, however, this effect has seldom been demonstrated on the synthesis of nanotubes. Part of the reason appears to be the greater challenges involved in fabricating nanotubes from nanowires because the nanowires available are normally too thick (~100 nm) to undergo smooth Kirkendall conversion to well-defined nanotubes. Recently, we succeeded in synthesizing ultrathin Zn nanowires down to a few nanometers by vapor transport deposition with the assistance of small molecules. For the Kirkendall conversion to ZnO nanotubes, the black Zn nanowire products were heated in air from 25 to 400°C by the rate of 15°C/min, and kept at 400°C for different periods of time (e.g., 0.5 h, 4 h) (Qiu and Yang, 2008).

Figure 2.16 presents TEM snapshots of the nanostructures in the course of transformation from Zn nanowires to ZnO nanotubes. The starting material is single-crystalline Zn nanowires. It is clearly seen that the products have a wire-like morphology with an average diameter of about 8 nm, although seemingly tangled and intertwined. After the Zn nanowires were heated in air at 400°C for 30 min, some bubble-like voids and hollow channels inside the nanowires can be seen in the TEM image shown in Fig. 2.16B. When the heating time was increased to 4 h, the bubble-like voids in the nanowires become depleted and at the same time, the hollow channels become more abundant, connoting the formation of nanotubes. The inner and outer diameters of the nanotubes are about 4 and 13 nm, respectively, with the latter being expanded relative to that of the original Zn nanowires. The HRTEM image of an as-formed ZnO nanotube shows that the nanowire is single crystalline. These fringes are equally spaced by 0.13 nm, which concurs well with the interplanar spacing of (0004) ZnO and alludes to the ZnO nanotube growth direction along [0001].

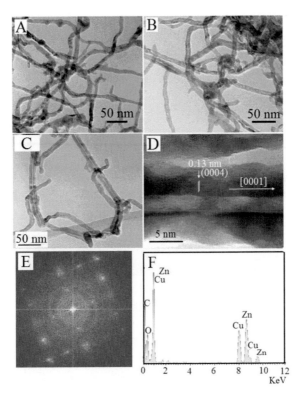

Figure 2.16 TEM images of Zn nanowires (A), Zn(void)-ZnO core–shell nanowires (B), and ZnO nanotubes (C); (D) HRTEM image of a single ZnO nanotube from (C); (E, F) Fast Fourier transform (FFT) pattern and EDS from (D).

In Figure 2.17, the formation of the ZnO nanotubes can be well explained by the Kirkendall effect. First, Zn nanowires are obtained in an O_2-free atmosphere by vapor transport deposition technique. The Zn nanowires are then heated in air from 25 to 400°C and kept at 400°C for 4 h. The Zn nanowire surfaces were first oxidized and coated with a layer of ZnO, leading to the formation a core–shell tubular nanostructure; internal voids are developed at the same time due to the Kirkendall effect. With the lapse of time, the Kirkendall voids continue to expand, leading to the final formation of the ZnO nanotubes. Because the solid-state cross-diffusion is a critical step, the Kirkendall approach is particularly useful for the synthesis of ultrathin nanotubes as we have demonstrated earlier.

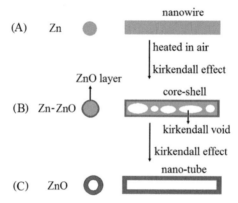

Figure 2.17 Schematic illustration of the ZnO nanotube formation process by the Kirkendall mechanism. (A) Zn nanowires; (B) Zn(void)-ZnO core–shell nanowire with internal Kirkendall voids; and (C) ZnO nanotube.

2.12 Layered Double Hydroxides Synthesis

Layered double hydroxides (LDHs) are a class of two-dimensional (2D) anionic clays made up of positively charged brucite-like host layers and exchangeable charge-balancing interlayer anions (Fig. 2.18a) that can be expressed as $[M^{2+}_{1-x}M^{3+}_{x}(OH)_2]_x^+ (A^{n-})_{x/n}$ mH_2O. Generally, LDH materials can be prepared by direct precipitation of mixed metal hydroxides in a solution. In theory, any divalent and trivalent metal ions with radii close to that of Mg^{2+} can constitute the host slabs of LDHs. However, transition

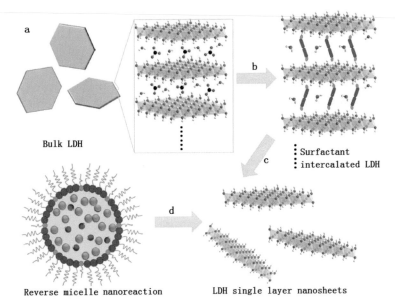

Figure 2.18 Schemes illustrating (a) the structure of carbonate-intercalated LDHs with the metal hydroxide octahedral layers stacked along the crystallographic c-axis; (b) the anion exchange process of LDH; (c) the delamination/exfoliation of LDH into single-layer nanosheets; (d) the bottom-up synthesis of LDH nanosheets in a reverse micelle.

metals, especially Fe^{3+}, are prone to form gel-like hydroxides at low pH values, making it difficult to synthesize transition-metals-based LDHs (TM-LDHs) by homogeneous precipitation. On the other hand, by using urea or hexamethylenetetramine as hydrolysis agents, highly crystallized TM-LDH microplates were hydrothermally synthesized in a recent study. The crucial point here is the progressive hydrolysis of urea or hexamethylenetetramine that makes the solution alkaline and induces homogenous nucleation and crystallization of LDH materials. In addition to the hydrothermal method, Ma et al. synthesized a series of TM-LDHs with a highly crystalline hexagonal microplatelet structure through topo-chemical transformation. The advantage of this method lies in the rational control of the transition-metal oxidation states in the hydroxide precursors. Recently, a facile direct electrochemical deposition method and microwave-assisted methods were also used to synthesize LDH-based materials, including NiCo, CoAl LDH,

ZnCo LDH, etc., greatly broadening the landscape of LDH synthesis. In addition to the nanoparticle or nanoplate structures, LDH-based core–shell, nanocone and nanoflower structures have also been reported. By studying the prototypes of ZnAl LDH and CoAl LDH, Forticaux et al. recently found that the growth of LDH was driven by screw dislocations. By controlling and maintaining a low precursor supersaturation, LDH nanoplates with well-defined morphologies could be synthesized. In contrast, the uncontrolled overgrowth led the LDH to develop a nanoflower morphology. Research on the applications of LDH materials continues to expand crossing many disciplines, such as catalysts, catalyst precursors, anion exchangers, and electro-active/photoactive materials. However, many of the applications are limited in scope because of the inaccessibility to the inner surfaces of the host layer. An effective solution to this problem would be to synthesize LDHs nanosheets with only a single highly anisotropic layer, which not only exhibit high surface area but also fully expose the electrochemically active sites. The fabrication of LDH nanosheets can be generally classified into bottom-up and top-down methods. The top-down method relies on the exfoliation/delamination of bulk LDH into single-layer nanosheets, which starts with and enabled by the ion-exchange intercalation involving an anionic surfactant to enlarge the brucite interlayer distance and weaken the brucite interlayer interaction (Fig. 2.18b). More often than not, the exfoliation/ delamination is carried out in a highly polar solvent, such as butanol, acrylates, CCl_4 and toluene, formamide, water and other solutions (Fig. 2.18c), which can solvate the hydrophobic tails of the intercalated anions. Therefore, several steps need to be taken to overcome the high interlayer charge density as mentioned earlier to achieve the full exfoliation/delamination of bulk LDH. Besides the top-down method, LDH nanosheets can also be prepared via the bottom-up synthesis. Typically, it begins with the introduction of coprecipitation of an aqueous precursor solution into an oil phase with a surfactant and a cosurfactant (Fig. 2.18d). The resulting reverse micelles act as nanoreactors, in which LDH single layers can be formed due to limited nutrients available in a confined space. Through such controlled nucleation, MgAl LDH, NiAl LDH, and CoAl LDH monolayer species were successfully synthesized. Notably, the control over the water and oil phases as well as the metal salts is stringent in order to realize the synthesis of LDH nanosheets by this method. Moreover, the successfully synthesized LDH nanosheets by the bottom-up strategy are still limited in both the production quantity and variety. So till now, the top-down fabrication is thus far the most widely developed and applied method for preparing LDH nanosheets. More recently, a new kind of LDH complexes composed of LDH nanosheets and 2D carbon materials, including GO and rGO as interlayer anions, were reported. Besides the high surface area, the 2D heterostructure LDH complexes also showed extremely high electrochemical activity due to the direct contact between catalytic metals and conductive carbons. TM-LDHs with intrinsic electrochemical activities, when further increased by designing suitable nanosheets structures, will be critical for the application of LDHs in energy-related processes, which will be discussed at some length in the following section (Long et al., 2016).

2.13 Binary-Nonmetal TMCs Synthesis

The growing concern about global warming, environmental pollution, and energy security has raised the demand for clean energy resources in place of fossil fuel. Cost-efficient generation of hydrogen from water-splitting through electrocatalysis holds tremendous promise for clean energy. Central to the electrocatalysis are efficient and robust electrocatalysts composed of earth-abundant elements, which are urgently needed for realizing low-cost and high-performance energy-conversion devices. Transition-metal compounds (TMCs) are a group of attractive noble-metal-free electrocatalysts for the hydrogen evolution reaction (HER). The incorporation of foreign nonmetal atoms to TMCs is a way of controllable disorder engineering and modification of electronic structure, and thus may realize the synergistic modulations of both activity and conductivity for efficient HER performance. In the last few years, the interest in binary-nonmetal TMCs as an efficient HER electrocatalyst is growing exponentially owing to their fascinating electronic structure and chemical properties.

Different synthetic routes would result in distinct morphology and structure of the materials and thus have pronounced impact on the HER activities. In general, two distinct strategies have been developed for the synthesis of binary-nonmetal TMCs: (1) foreign nonmetal atoms doping and (2) one-step preparation via mixed precursors. Foreign nonmetal atoms doping based on the parent TMC materials is the most common synthetic strategy to synthesize binary-nonmetal TMCs. According to the type of doped elements, this versatile synthetic strategy can be further divided into phosphorization, sulfuration, selenization, and nitridation. The solid/gas-phase doping is a general approach for the synthesis of binary-nonmetal TMCs. It is noteworthy that post-treatment of tail gas is extremely necessary. For example, selenium and sulfur powders are general reagents used for the selenization and sulfuration reactions, respectively. The in situ formed Se vapor and SeO_2 in the selenization process, and S vapor and SO_2 in the sulfuration process, are toxic. PH_3 is an efficient and active precursor in phosphorization but is extremely toxic and lethal even at a few ppm. Although less toxic phosphorization reactants such as hypophosphites NaH_2PO_2, red phosphorus powders are widely utilized, post-treatment of tail gas is still extremely necessary to adsorb in situ generated PH_3 and phosphorus vapor. As shown in Fig. 2.19A, P was incorporated into the CoS_2 structure via a solid/gas-phase reaction at 400°C by using NaH_2PO_2 as the precursor, which can in situ form PH_3 (Hu et al., 2017).

One-step synthetic approaches that enable the formation of stoichiometric compositions are significant synthetic strategy. Xiang et al. reported the first synthesis of monolayer $WS_{2(1-x)}Se_{2x}$ triangular domains using a chemical vapor deposition route. Recently, Jin et al. used the pyrolysis method to synthesize pyrite-type cobalt phoshosulfide where thiophosphate precursor was first prepared and located in the upstream side and the cobalt precursor substrates were placed in the downstream side. As shown in Fig. 2.19B, He et al. designed a district pyrolysis system to synthesize $WS_{2(1-x)}Se_{2x}$ compounds, in which the mixed S and Se powders were located in front zone of the quartz tube with a lower temperature to generate S and Se vapor and the WO_3 nanowires precursor were placed in back zone of the tube with higher temperature to facilitate the sulfuration and selenization reactions. The crucial point here is the progressive pyrolysis of S, Se, and P molecule precursors into atoms that possess high activity. The use of pyrolysis synthetic methods currently available for the preparation of binary-nonmetal TMCs guarantees certain positive outcomes but suffer from major drawbacks as well. For example, because the chemical reactivity of Se is lower than

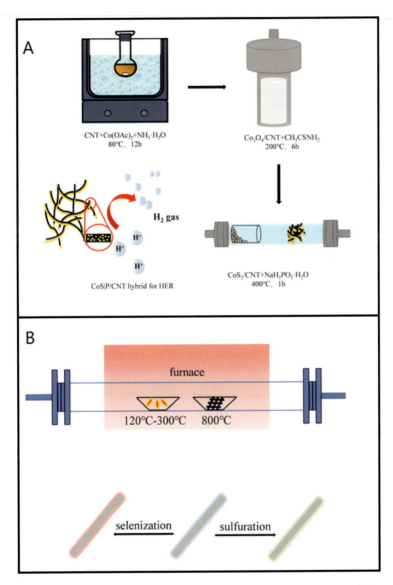

Figure 2.19 (A) Schematic presentation of the foreign nonmetal atoms doping to prepare the CoS|P/CNT hybrid materials; (B) schematic illustration of the one-step synthesis for the $WS_{2(1-x)}Se_{2x}$ nanotubes.

that of S, the selenization condition should be changed compared with the sulfuration condition by increasing the temperature to enhance the chemical reactivity or separating the reaction region of selenization and sulfuration. Therefore, in order to obtain products with stoichiometric compositions, complete reaction under high temperature and long reaction time is crucial. Gaudin et al. obtained $Ag_2Ti_2P_2S_{11}$ single crystals by heating the stoichiometrically proportional reactants mixture at high temperature of 873 K for 10 days. The high-temperature solid-state synthesis method was also used by Mukherjee et al. for preparing $FePS_3$ crystals after quartz tube heating at 973 K for 6 days. To synthesize $MoS_{2(1-x)}Se_{2x}$ crystals with different x values, Sampath et al. mixed stoichiometric proportions of molybdenum, sulfur, and selenium powders and heated the mixture in quartz tube at 1073 K for 3 days. In addition to the high-temperature solid-state methods, facile hydrothermal/solvothermal method and hot-injection method were also used to synthesize the binary-nonmetal $MoS_{2(1-x)}Se_{2x}$ materials. For example, Yan et al. reported the preparation of S-doped $MoSe_2$ nanosheets by reacting Mo-oleylamine and Se-oleylamine-dodecanethiol mixtures, in which dodecanethiol as the surfactant can not only promote the dissolution of the selenium in oleyamine but also lower the surface energy of the basal edges, resulting in nanosheets with large quantities of exposed edge sites and defects. These studies greatly broadened the landscape of the binary-nonmetal TMCs synthesis.

2.14 Fully Inorganic Trihalide Perovskite Nanocrystals Synthesis

Fully inorganic trihalide perovskite nanocrystals (NCs) are emerging as a new class of superstar semiconductors with excellent optoelectronic properties and great potential for a broad range of applications in lighting, lasing, photon detection, and photovoltaics. Over the past 2 years, two classes of preparation procedures have been developed to fabricate all-inorganic perovskites NCs, namely, the solution chemical process and chemical vapor deposition (CVD). The CVD method can achieve epitaxial growth of single-crystal $CsPbX_3$ (X = Cl, Br, I) nanowires on crystalline substrates. However, the relatively higher reaction temperature and difficulty in engineering morphology hinder the CVD procedure in the preparation of perovskite NCs with various shapes, sizes, and dimensions. The widely used solution approaches offer an effective route for the fabrication of uniform perovskites NCs, including 0D QDs and nanocubes, 1D nanowires and nanorods, and 2D nanoplatelets and nanosheets. In the following section, several representative procedures are briefly discussed: hot-injection route, RT ligand-mediated reprecipitation, RT supersaturated recrystallization process, droplet-based microfluidic technology, and solvothermal method. Then, some strategies for controlling morphology and size of perovskites NCs are summarized, in which key factors for modulating shape and size are discussed (He et al., 2017).

2.14.1 High-Temperature Hot-Injection Route

In 2015, following the traditional hot-injection procedure, which is commonly used for the synthesis of metal chalcogenide or fluoride NCs, Kovalenko et al. pioneered the fabrication of $CsPbX_3$ (X = Cl, Br, I, and their mixtures) NCs. Subsequently, several groups used this approach to colloidal NCs of $CsPbX_3$ with different halide compositions, shapes, and sizes. Generally, the formation of $CsPbX_3$ NCs is performed by the swift injection of cesium oleate into an octadecene solution containing PbX_2, oleic acid (OA), and oleylamine (OAm) at high temperatures (for example, 140–200°C). The mixture of OAm and OA can solubilize PbX_2 and colloidally stabilizes the as-obtained perovskite NCs. Hence, both carboxylic acid and amines are indispensable for the successful synthesis of the perovskite NCs. Cesium oleate can be replaced by other Cs-containing organometallic compounds such as cesium stearate and cesium acetate. The use of a more soluble cesium acetate precursor resulted in enhanced processability and smaller sizes without sacrificing optoelectronic properties. As for the synthesis of tin-based halide perovskite (such as $CsSnX_3$) NCs, the tin precursor was used to form the complex between SnX_2 and tri-n-octylphosphine (as both the mildly reducing and coordinating solvent) instead of SnX_2.

The formation reaction of $CsPbX_3$ NCs can be expressed as

$$2\text{Cs-oleate} + 3PbX_2 \rightarrow 2CsPbX_3 \text{ NCs} + Pb(\text{oleate})_2 \quad (2.6)$$

Herein, PbX_2 is the sole source of X^- ions and hence one-third of Pb^{2+} will be spent for the formation of lead oleate as byproduct. Accordingly, apart from the reaction temperature, lead-to-cesium molar ratio (R) is another key parameter to obtain CsPbX$_3$ perovskite NCs. The value of $R \geq 1.5$ is generally required. For example, the R ratio was set as 3.76 in the first investigation by Kovalenko et al.

As illustrated in Eq. (2.6), the formation reaction of CsPbX$_3$ NCs is conducted with excess halide, and cesium oleate is the limiting reagent. Before isolation and purification, aside from CsPbX$_3$ NCs, lead oleate, oleylammonium halide, OA, and OAm coexist in the post-reaction system, which all serve as surface-binding species. Roo et al. found that ligands binding to the NC surface easily vanished during the isolation and purification procedures because CsPbX$_3$ species are ionic in nature and the interactions with capping ligands are also more ionic and labile. However, adding small amounts of excess OA and OAm before precipitation can preserve the colloidal integrity and PL of the NCs. In addition, they also demonstrated that the presence of excess amines in the solution after purification can result in high QY owing to the improved binding of the carboxylic acid.

OA and long-chain alkyl-amines (dodecylamine, oleyl amine, and n-octylamine) can break the crystal's inherent cubic symmetry and guide the 2D growth of CsPbBr$_3$. With the aid of OA and long-chain alkyl-amines as the growth-directing soft templates, CsPbBr$_3$ nanosheets with a thickness of about 3.3 nm and an edge length of about 1 μm have successfully been fabricated. After investigating the impacts of ether solvents (i.e., ethylene glycol dibutyl ether, diethylene glycol dibutyl ether, and tetraethyleneglycol dibutyl ether) with different polarity on the nucleation and growth of CsPbBr$_3$ QDs, Li et al. found that the size and morphology of QDs could be better controlled in the solvent with low polarity. Consequently, the reaction medium is also a key parameter to the hot-injection synthesis of CsPbX$_3$ NCs.

Currently, as the most widely adopted procedure for the fabrication of CsPbX$_3$ NCs, the high-temperature hot-injection approach is a very promising methodology to achieve high-quality perovskite NCs. In this procedure, the non-coordination solvent used is 1-octadecene, while OA and OAm are chosen as the ligands for stabilizing the resulting NCs. The formation of CsPbX$_3$ NCs was found to occur through two separate stages: seed-mediated nucleation followed by further growth via oriented attachment and self-assembly. As known to all, size uniformity is the most critical factor in exploiting the size-dependent properties of NCs. Due to the separation of the nucleation and growth stages in the hot-injection synthesis, a high degree of monodispersity can be achieved without using further size-selective techniques. However, it is worth noting that the objective perovskite NCs with desired phase, shape, and size depend to a great extent on the isolation and purification of post-reaction mixtures, which is commonly believed to be a tough step.

2.14.2   Room-Temperature Coprecipitation Method

The reaction temperature in the hot-injection method is uncontrollable upon the injection of cold precursor solution, resulting in low batch-to-batch reproducibility. This problem will become more prominent in the large-scale production, and it should be resolved by developing other alternative procedures. In this regard, the following RT synthetic strategies have been developed, which can be easily scaled up by scaling up each of the reaction components.

2.14.2.1   Ligand-mediated reprecipitation method

Akkerman et al. most recently reported an RT route in which the nucleation and growth of $CsPbBr_3$ nanoplatelets are triggered by the injection of acetone into a precursor mixture of $PbBr_2$, Cs-oleate, OAm, OA, and HBr, which would remain unreactive otherwise. Other polar solvents such as isopropanol and ethanol were able to ensure the nucleation and growth of halide perovskite NCs but were not as effective as acetone in shape manipulation. They claimed that acetone most likely destabilizes the complexes of Cs^+ and Pb^{2+} ions with the various molecules in solution and, therefore, initiates the nucleation of the particles.

In 2015, Zhong et al. developed an RT ligand-assisted reprecipitation technique for the fabrication of $CH_3NH_3PbX_3$ (X = Cl, Br, I) QDs. Subsequently, Deng et al. utilized this strategy for the shape-controlled synthesis of all-inorganic $CsPbX_3$ (X = Cl^-, Br^-, I^-) perovskite NCs. This strategy was conducted by mixing a precursor solution in polar solvent (such as N,N-dimethylformamide (DMF), tetrahydrofuran, and dimethyl sulfoxide) with a nonpolar solvent (such as toluene and hexane) at RT. The precursor solution consists of a polar solvent, PbX$_2$, Cs-oleate, octadecene, organic acid, and amine ligands.

DMF as reaction medium can severely degrade and even dissolve CsPbX$_3$ (X = Cl, Br, I) NCs, which results in a decrease in production yield. Considering this drawback, Wei et al. developed an RT homogeneous reaction procedure for preparing CsPbBr$_3$ perovskite QDs in the atmospheric environment, which can be readily extended to gram-scale preparation. CsPbBr$_3$ QDs could be obtained by mixing Cs$^+$ and Pb^{2+} solutions with a Br$^-$ precursor solution in a variety of nonpolar organic solvents (such as chloroform, dichloromethane, xylene, or hexane and toluene). Cesium and lead fatty acid (such as butyric acid, hexanoic acid, octanoic acid, decanoic acid, or myristic acid, and OA) salts and quaternary ammonium bromides (such as tetrabutyl ammonium bromide and tetraoctyl ammonium bromide) were used as the precursors of Cs$^+$, Pb^{2+}, and Br$^-$, respectively. Aside from the complex, the excess fatty acids with different chain lengths (C4–C18) acted as capping agents. They found that the PL QY of as-prepared CsPbBr$_3$ NCs exceeds 80% in the case of a myristic acid or OA as the capping agent, while the QY of the as-obtained butyric-acid-capped CsPbBr$_3$ NCs reached as low as 20–30%, probably due to a weaker bonding strength. Wei et al. also investigated the possibility of obtaining CsPbBr$_3$ NCs in the presence of some organic amines (such as butylamine, hexylamine, octylamine, OAm, and trioctylamine), 1-dodecanethiol, trioctylphosphine, trioctylphosphine oxide, and triphenylphosphine oxide as capping agents. Unfortunately, no objective CsPbBr$_3$ NCs were formed in the presence of these strong capping agents since the nucleation of NCs was completely inhibited at RT. All these results demonstrated that fatty acid plays a crucial role in the RT homogeneous route to CsPbBr$_3$ NCs, which is consistent with our achievement in the preparation of polynary fluoride NCs.

2.14.2.2 Supersaturated recrystallization process

Inspired by the report from Zhong et al., Zeng et al. designed an RT supersaturated recrystallization procedure to fabricate CsPbX$_3$ (X = Cl, Br, I) QDs, which can be finished within a few seconds and is free from conventional heating, protective atmosphere, and injection operation. Due to the huge differences of solubilities for Cs$^+$ and Pb^{2+} in DMF (polar solvent) and toluene (nonpolar solvent) (as large as more than six orders of magnitude), the transfer from DMF to toluene produces a highly supersaturated state immediately and then induces rapid recrystallization under vigorous stirring according to Eq. (2.7):

$$Cs^+ + Pb^{2+} + 3X^- \rightarrow CsPbX_3 \quad (2.7)$$

In this RT recrystallization process (Eq. 2.7), the reaction was very intense and rapid. It is, therefore, hard to kinetically control the QD growth, resulting in increasing the difficulty in tuning the shape and size of QDs.

2.14.3 Droplet-Based Microfluidic Approach

The aforementioned procedures focused on a conventional batch or flask-based reactions. In the standard batch systems, the nucleation rates for CsPbX$_3$ NCs are comparable to (or even faster than) the speed of homogeneous mixing of reagents and heat transfer, which limits the understanding of the parameters governing the nucleation of these NCs. In this regard, Lignos et al. developed a droplet-based microfluidic platform for studying growth kinetics and optimizing the synthetic parameters of colloidal CsPbX$_3$ NCs. It was found that this system allows the rigorous and rapid mapping of the reaction parameters, including molar ratios of lead, cesium, and halide precursors, reaction temperatures, and reaction times. Moreover, it can perform ultrafast kinetic measurements and reaction optimization through the combination of online absorbance and fluorescence detections and the fast mixing of reagents. It allows for precise tuning of the chemical payload of the formed droplets through continuously changing the ratio between Pb^{2+} and Cs$^+$ sources as well as the ratio between halides. In particular, their investigation presented the refined parameters that can be applied to the batch syntheses, and also gave unique insights into the early stages of nucleation and growth within the initial 0.1–5 s of CsPbX$_3$ NCs. Compared to batch or flask-based synthetic approaches, the droplet-based microfluidic strategy can also enormously save reagent usage and screening times.

2.14.4 Solvothermal Method

Solvothermal synthesis not only offers great advantages of maneuverability, versatility, and low cost, but also can yield high crystallinity and well-defined NCs under modest temperatures and increased pressures. Therefore, this facile solvothermal route was introduced to fabricate high-quality $CsSnX_3$ (X = Cl, Br, and I) quantum rods for the first time by Chen et al. In their investigation, SnX_2 and Cs-oleate acted as Sn^{2+}, X^-, and Cs^+ sources, respectively. A mixture of diethylenetriamine, 1-octadecene, OA, and OAm was used as reaction medium, while trioctylphosphine oxide was utilized as complexing agents for Sn^{2+} and Cs^+. This one-pot fabrication was performed in a sealed Teflon-lined autoclave at 180°C, in which the reaction temperature is lower than the boiling points of the solvents used. According to the solvothermal chemistry, strictly speaking, this route cannot be ascribed to the solvothermal method. Furthermore, although they claimed that the optical characteristics of the as-obtained $CsSnX_3$ quantum rods can be changed through solvothermal without an annealing process, the effect of various synthetic parameters (especially reaction temperature) on the phase purity, shape, and morphology of the products is not revealed.

2.15 Conclusion

In summary, in this chapter, several representative clean energy materials are covered with respect to their synthesis methods, which can be used to engineer their structures and morphologies. First, 0D material CDs could be synthesized by top-down methods or bottom-up methods. Second, 1D materials could be synthesized by several methods. For example, copper sulfide nanowire arrays growing on a copper substrate, Ag_2S nanowires on silver substrates, and α-Fe_2O_3 nanobelt/nanowire arrays on iron substrates were synthesized by the GS approach; $Cu(OH)_2$ nanoribbon arrays on copper foil and ZnO nanofibers aligned on the zinc substrate were synthesized by the SS reaction approach; to modify the surface of nanowire, Cu_2S/Au core/sheath nanowire arrays and Cu_2S/polypyrrole core/sheath nanowire arrays were obtained via the redox reactions of galvanic cells and an interfacial polymerization technique; ultrathin Bi_2O_3

nanowire and ultrathin ZnO tetrapods could be prepared by an oxidative metal vapor transport deposition technique. Third, 2D material LDHs nanosheets can be generally fabricated by bottom-up and top-down methods. Fourth, three-dimensional (3D) material binary-nonmetal TMCs, which are a group of attractive noble-metal-free electrocatalysts for the HER, could be synthesized by foreign nonmetal atoms doping or one-step preparation via mixed precursors. Finally, fully inorganic trihalide perovskite NCs with excellent optoelectronic properties could be prepared by the solution chemical process and CVD.

There are simply too many new materials for clean energy applications to be listed here exhaustively. The representative materials discussed in this chapter can hopefully give the readers inspiration and entry points for their own exploration.

Questions

1. Please write five new clean energy material preparation methods.
2. What are the growth mechanisms of certain nanowires?
3. Can you design a method to fabricate cube-shaped nanoparticles?
4. Can you design a method for preparing ZnO nanotubes?
5. How to prepare nanowires with a heterostructure?

References

He X., Qiu Y., and Yang S., "Fully-inorganic trihalide perovskite nanocrystals: A new research frontier of optoelectronic materials." *Adv. Mater.*, **29**(32), 1700775 (2017).

Hu J., Zhang C., Meng X., Lin H., Hu C., Long X., and Yang S., "Hydrogen evolution electrocatalysis with binary-nonmetal transition metal compounds." *J. Mater. Chem. A*, **5**(13), 5995–6012 (2017).

Long X., Wang Z., Xiao S., An Y., and Yang S. "Transition metal based layered double hydroxides tailored for energy conversion and storage." *Mater. Today*, **19**(4), 213–226 (2016).

Qiu Y. and Yang S. "Kirkendall approach to the fabrication of ultrathin ZnO nanotubes with high resistive sensitivity to humidity." *Nanotechnology*, **19**(26), 265606 (2008).

Qiu Y. and Yang S. "ZnO nano-tetrapods: controlled vapor phase synthesis and novel application for humidity sensing." *Adv. Funct. Mater.*, **17**(8), 1345–1352 (2007).

Qiu Y., Liu D., Yang J., and Yang S. "Controlled synthesis of bismuth oxide nanowires by an oxidative metal vapor transport deposition technique." *Adv. Mater.*, **18**, 2604–2608 (2006).

Qiu Y., Yang M., Fan H., Zuo Y., Shao Y., Xu Y., Yang X., and Yang S. "Nanowires of α- and β-Bi_2O_3: phase-selective synthesis and application in photocatalysis." *CrystEngComm*, **13**(6), 1843–1850 (2011).

Yuan F., Li S., Fan Z., Meng X., Fan L., and Yang S. "Shining carbon dots: synthesis and biomedical and optoelectronic applications." *Nano Today*, **11**(5), 556–586 (2016).

Yuan T., Meng T., He P., Shi Y., Li Y., Li X., Fan L., and Yang S. "Carbon quantum dots: an emerging material for optoelectronic applications." *J. Mater. Chem. C*, **7**(23), 6820–6835 (2019).

Zhang Q., Zhang K., Xu D., Yang G., Huang H., Nie F., Liu C., and Yang S. "CuO nanostructures: synthesis, characterization, growth mechanisms, fundamental properties, and applications." *Progress Mater. Sci.*, **60C**, 208–337 (2014).

Zhang W. and Yang S. "In situ fabrication of inorganic nanowire arrays grown from and aligned on metal substrates." *Acc. Chem. Res.*, **42**(10), 1617–1627 (2009).

Chapter 3

Interface Engineering for Perovskite Solar Cells

For the design and fabrication of functional devices, the importance of material interfaces can never be overestimated as the famous saying goes, "the interface is the device." In this chapter, we will use the rapidly rising perovskite solar cells (PSCs) as a case to illustrate the rich facets of material interface engineering.

3.1 Introduction of Perovskite Solar Cells

Perovskite solar cells can be fabricated into either a conventional n-i-p or an inverted p-i-n structure. The conventional n-i-p structure PSCs generally consist of a transparent conductive oxide (TCO) substrate, an electron transport layer (ETL), a light-harvesting layer (organometal halide perovskites), a hole transport layer (HTL), and a metal electrode (Au, Ag, etc.). When an ETL, usually n-type wide bandgap oxide semiconductor (TiO_2, ZnO, SnO_2, etc.), is deposited on the TCO substrate, the conductive substrate is a negative electrode. In contrast to the n-i-p structure device, the inverted p-i-n structure device is constructed by first depositing an HTL layer on TCO, which plays the role of positive electrode when the cell is working. Since the inverted p-i-n structure PSCs can be produced at a relatively low temperature, most flexible PSCs are based on this architecture.

Materials and Interfaces for Clean Energy
Shihe Yang and Yongfu Qiu
Copyright © 2022 Jenny Stanford Publishing Pte. Ltd.
ISBN 978-981-4877-66-4 (Hardcover), 978-1-003-14223-2 (eBook)
www.jennystanford.com

3.2 Importance of Interfaces in Planar p-i-n PSCs

Perovskite semiconductors are quite different from the organic semiconductors for photovoltaic applications whose recombination dynamics are usually dominated by excitons. The free-carrier model is more suitable than the exciton model for interpreting the properties of perovskite, wherein charge transport and separation take place more like in a heterojunction solar cell. According to the Anderson model and thermal equilibrium theory, when two types of semiconductors contact directly, the free carriers will diffuse toward each other spontaneously, equalizing the Fermi energy level and yielding a charge-depletion region with a built-in electric field known as a junction. The junction is mainly based on the thermal equilibrium theory. When light is absorbed by a solar cell with a junction, the thermal equilibrium will be broken, and charge extraction and transport processes occur. The photoinduced carriers have to transport across the interfaces in the cell, and charge loss usually occurs due to interfacial defects.

In PSCs, charge extraction occurs at the interfaces, which may be particularly subject to recombination mainly due to any possible interfacial defects and the associated specific charge distributions. In the planar p-i-n PSCs, the hole transport material, perovskite absorber, electron transport material, and the metal electrode are sequentially deposited onto the transparent conducting substrate (fluorine-doped tin oxide (FTO) or indium tin oxide (ITO)), to develop an entire solar cell. The charge is extracted at the HTM/perovskite and the perovskite/ ETM interfaces and is collected through the front and back contact interfaces by vertical and lateral transport. In order to achieve highly efficient PSCs, numerous works have mainly focused on the design of new materials and device structures, and the control over perovskite film quality. It appears impending now that interface engineering still has a big, untapped role to play for enhancing both efficiency and stability of PSCs. Among several functions of the interfacial layers in PSCs, we will discuss their contributions to energy-level alignment, charge dynamics, and trap passivation in the following subsections.

3.2.1 Energy-Level Alignment

Energy-level alignment at the interfaces is critical for the optimization of the solar cell device. Appropriate energy-level tailoring can increase open-circuit voltage V_{oc} and facilitate charge transfer and extraction, which contribute to increasing short-circuit current density J_{sc} and fill factor FF. The two important surfaces for energy-level alignment are formed by perovskite with HTM or ETM. To facilitate the carrier transfer, the lowest unoccupied molecular orbital (LUMO) (or conduction band minimum (CBM)) of perovskite should be higher than that of ETM and the highest occupied molecular orbital (HOMO) (or valence band maximum (VBM)) should be lower than that of HTM. The band offsets of HTM/perovskite and ETM/perovskite are a key determinant of the carrier recombination at the relevant interfaces. Empirically, a band offset of around 0.2 eV is necessary to ensure an efficient charge extraction at the perovskite/ETM interfaces. And theoretical analysis by Murata an Minemoto also suggested that the band offset of HTM and perovskite should also be within 0.2 eV to make an efficient PSC device. The energy levels for the commonly used materials in planar p-i-n PSCs are shown in Fig. 3.2. The work functions of adjacent layers in planar p-i-n PSCs, which should match well with each other, can be effectively tailored by interface engineering.

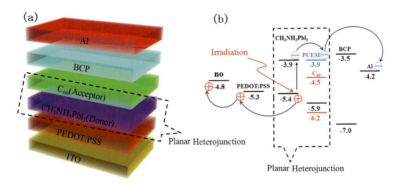

Figure 3.1 (a) Device configuration of the first planar p-i-n perovskite solar cell. (b) Respective energy level and charge transport diagram. (Bai et al., 2018).

52 | *Interface Engineering for Perovskite Solar Cells*

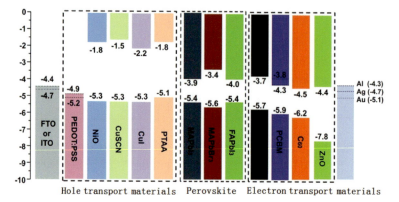

Figure 3.2 Schematic energy-level diagram of commonly used materials in planar p-i-n PSCs.

Band bending refers to the change in the monotonic band edge of a semiconductor near a junction due to the energy offset with respect to its junction partner. Such a band-bending effect was also observed at the interface between perovskite and transporting layers. Kahn et al. measured the band bending at the interface between 2,2′,7,7′-tetrakis[N,N-di(4methoxyphenyl)amino]-9,9′-spirobifluorene (Spiro-OMeTAD) and perovskite. They found that Spiro-OMeTAD has a lower ionization energy and thus forms a non-optimum energy-level alignment with perovskite. This would result in band bending upward toward the interface with perovskite, which could limit the V_{oc} of solar cell devices. For the same reason, they also suggested that establishing an optimized energy-level alignment should facilitate hole extraction. Interface engineering is expected to be beneficial for tuning the interfacial energy level and decreasing the energy gap between perovskite and the transporting material, further minimizing V_{oc} losses at the junction (Bai et al., 2018; Chen and Yang, 2019).

3.2.2 Charge Dynamics

The charge dynamics happened at the interface, including charge extraction, charge transfer, and charge recombination. The charge extraction is a quick process. It is found that within hundreds of picoseconds, the photogenerated electrons and holes could be extracted to the ETM and HTM. Compared with the lifetime of free carriers, which is at the timescale of nanoseconds, the extraction process is ultrafast. However, some works demonstrated that the remaining PbI_2 at interface may slow the charge extraction process. The surface modification was developed as an efficient method to promote the ultrafast interfacial charge extraction. In one of our former works, we used carbon quantum dots, which acted as superfast electron tunnel, to modify the interface between ETM and perovskite, resulting in an accelerated charge extraction process, which increased the extraction rate from ≈300 ps to just around 100 ps. Compared with the charge extraction process, charge transfer and recombination at the interface have a much stronger impact on the device performance like V_{oc} and J_{sc}. Interfacial structural and electronic mismatches usually act as the energy barriers for charge transport and charge recombination. Interface engineering is needed to eliminate this effect. In the work of Ogomi et al., they inserted the $HOCO-R-NH_3^+I^-$ group between ETL/perovskite interface, which greatly suppressed the charge recombination from picosecond to tens of microsecond. And in Chen's work, the development of hybrid interfaces also significantly increased the recombination resistance and suppressed the recombination process at the p-i-n PVSCs. Impedance spectroscopy (IS) is one of the few rare techniques that can be used to investigate the recombination in PSC devices. Bisquert et al. developed a general model to analyze various IS spectral features and established the relationship between frequency region and recombination resistance under different conditions. Typically, three arc features are observed in a PSC at low, intermediate, and high frequencies. The high-frequency component (i.e., ≈10^5 Hz) is attributed to the selective contact, while the intermediate (i.e., ≈10^3 Hz) and low frequencies (i.e., ≈1 Hz) are attributed to chemical capacitance and intrinsic properties of perovskites, respectively. Especially intriguing is the low-frequency arc, which has never been observed in other solid-state solar cells. Due to the slow timescale, the low-frequency arc was interpreted as arising from ion migration in the perovskite film under study. Detailed discussion of such an ion migration process will be given below. Besides, the IS measurements were also used to study other charge dynamic processes, such as surface polarization, charge collection, hysteresis phenomenon, etc.

3.2.3 Trap Passivation

Trap state at the perovskite surface and interfaces can lead to the charge accumulation and recombination losses in the device, and it had been observed that the passivation of trap states can eliminate the hysteresis phenomenon. In the work of Zhu et al., they found direct evidence that the hole traps exist on the surface of MAPbI$_3$ thin film. These hole traps were resulted from the under-coordinated halide anions on the crystal surface, which will lead to a significant charge accumulation of the perovskite/HTM interface. Snaith et al. introduced iodopentafluorobenezene to passivate the perovskite surface hole trap states, which strongly inhibited the charge recombination. Similar efforts had also been done by passivating the hole trap states with Lewis base. Besides the perovskite absorber, the trap states also existed at other interfaces; interface engineering had been applied to eliminate the trap states at the interfaces to reduce the recombination and charge-transfer barrier.

3.2.4 Ion Migration

Migration of the constituent ions (e.g., MA$^+$, Pb^{2+}, I$^-$) in organic–inorganic hybrid perovskites has received significant attention with respect to their critical roles in the device interface, hysteresis, and degradation in PSC. Numerous reports showed that the ions in perovskite materials (e.g., MA$^+$ or I$^-$) have low activation barriers and modest ionic diffusion coefficients to move within perovskite devices, especially when subjected to external bias or under light illumination. The ion migration was initially noticed through the anomalous hysteresis in the PSCs and subsequently found to be more general in explaining phenomena such as light-soaking effect and halide redistribution. Furthermore, ionic migration has adverse effects on the stability of PSCs because the migrating iodide ions can react with metal electrodes, resulting in the degradation of PSCs. Interface engineering is needed to control the ion migration and improve the efficiency and stability of the PSCs (Zhang et al., 2019).

3.3 Interface Engineering of Conventional n-i-p Structure PSCs

3.3.1 From Dye-Sensitized Solar Cells to PSCs

As mentioned earlier, organometal halide perovskites were first employed as the sensitizer in dye-sensitized solar cells (DSCs), which typically comprise mesoporous n-type oxide semiconductor (TiO_2, ZnO films composed by sintered nanocrystallines) as electron conductor, a dye as light absorber, a redox shuttle for dye regeneration, and a counter electrode to collect electrons and reduce positive charges generated through the cell. Before the introduction of organometal halide perovskites, the highest power conversion efficiency (PCE) of DSCs was 12.3%, obtained by using cobalt-based electrolytes together with the development of donor-π-acceptor porphyrin dyes. However, the liquid electrolyte-based DSCs lagged behind other photovoltaic technologies (polycrystalline silicon, CIGS, CdTe, etc.). They failed to achieve their theoretically attainable efficiency because of the loss in fill factor due to the large series resistance of the semiconductor nanocrystalline electrodes, loss in potential due to the overpotential required to drive the multitude of charge-transfer processes in iodine-based redox, and loss in photocurrent due to the low light absorption efficiency of the available dyes. For the conventional DSCs, oxide semiconductor photoanode needs to possess an extremely large surface to multiply the available area for dye anchoring. Even for dyes with high light absorption coefficients, the interfacial area has to be almost thousand-fold larger than that of a flat film, because only monolayer of dye sensitization can ensure sufficient electron transfer from dye to oxide photoanode. As a consequence, the mesostructured photoanode needs to be relatively thick, giving rise to the increase in the transport way and thus the recombination probability of the extracted electrons, since the photoanode constructed by sintered nanocrystallines can induce localized states below the conduction band (CB), which can trap and slow down the transported electrons. The obvious solution to this conundrum is to develop absorber materials that possess high molar extinction coefficients and perform effectively when more than a monolayer is deposited, such as semiconductor quantum dots (CdS, CdSe, ZnS, ZnSe, PbS, Sb_2S_3, etc.) and the organometal halide perovskites. Another way to tackle this issue is to construct photoanode with single or quasi-single crystals, like one-dimensional (1D) nanorod/wire/tube arrays, hierarchical tree-like three-dimensional (3D) nanostructures, or the mesoporous single crystals.

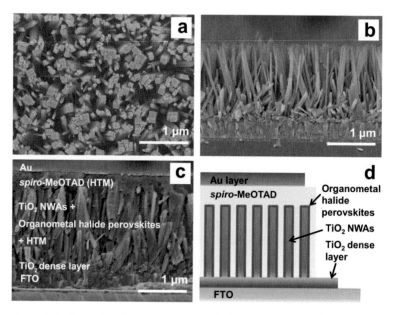

Figure 3.3 Top (a) and cross-sectional (b) view SEM images of TiO_2 NWAs synthesized on FTO substrate. (c) Cross-sectional SEM image of the TiO_2 NWAs/organometal halide perovskite/spiro-MeOTAD hybrid photovoltaic cell. (d) Schematic illustration of the solar cell. (Qiu and Yang, 2019).

Preceding solar cells based on organometal halide perovskites, researchers developed CdS/CdSe quantum dots, ZnSe/CdSe/ZnSe quantum well sensitizers for light harvesting, and a two-layer-structured photoanode composed of 1D ZnO nanowire arrays and ZnO tetrapod network, which exhibited high photoelectric conversion efficiencies. Furthermore, in the CdS/CdSe quantum dots and ZnSe/CdSe/ZnSe quantum well sensitizer systems, they found a double transport channel for the photogenerated charges by IS analysis, which is common in the PSCs but was new in the DSCs or quantum dot SCs at that time. However, the new-type inorganic light absorbers and the two-layer-structured photoanode could not be implemented in the solid-state DSCs until 1D TiO$_2$ nanowire arrays (NWAs) were introduced for the photoanode and organometal halide perovskite (in fact, among the first mixed perovskites used in perovskite solar cells) for light absorber. Such a device structure is shown in Fig. 3.3. The reason can be ascribed to the improvements in the electron transport and interface between TCO and oxide semiconductor photoanode induced by the 1D TiO$_2$ nanowire arrays, which will be discussed in detail in Section 3.3.2.

3.3.2 Interface between TCO and Oxide Semiconductor ETL

For the conventional n-i-p structure PSCs, one should deposit a compact TiO$_2$ layer, which is named hole-blocking layer before the deposition of ETL to avoid the direct connection of the TCO substrate and perovskite layer, especially for the TiO$_2$-based mesostructured PSCs. The TiO$_2$ compact layer played a curial role to achieve effective PCE in the earlier stage of PSC study. For the DSCs, where the mesostructured TiO$_2$ layer is directly deposited on TCO substrates, this TiO$_2$ compact layer for hole blocking is not necessary because of the very slow reaction rate of I$_3^-$ to 3I$^-$ at the interfaces between liquid iodide-based electrolyte and TCO substrate or TiO$_2$ film. Nevertheless, the PSC devices produced from the protocol copy of DSCs with simple modifications encountered short-circuit problem, which was mainly attributed to the severe recombination of the photogenerated electrons and holes at the TCO/perovskite interfaces, caused by the absence of compact layer for hole blocking. Another factor resulting in the failure of the PSC devices was the film thickness of the mesostructured ETL. Generally, the thickness of the mesostructured oxide semiconductor ETL of solid-state DSCs was typically in the range of 1–2 μm, which was too large for the PSC devices, giving rise to the huge resistance for electron transport and thus the very low PCE values. The issues of short-circuit and low PCE values continued for almost 3 years until researchers introduced the TiO$_2$ nanorod arrays to the device fabrication in 2012. The TiO$_2$ nanorods with single crystal were directly grown on the TCO substrate, which can provide much faster electron transport features compared to the mesoscopic TiO$_2$ film composed of nanoparticles, thus reducing the resistance for the electron transport although the film thickness is ~1.5 μm (Fig. 3.3b). Most importantly, they introduced a TiO$_2$ seed layer produced from high-temperature sintering of TiO$_2$ sol to grow TiO$_2$ nanorod arrays (Fig. 3.3c,d). This TiO$_2$ seed layer was compact, which can play the role of hole blocking (Fig. 3.3d). As a result, the PSC device fabricated by combing TiO$_2$ seed layer as hole-blocking layer, TiO$_2$ nanorod arrays as ETL, MAPbI$_3$/MAPbI$_2$Br as light absorber, spiro-OMeTAD as HTL, and Au as black electrode achieved a PCE value of ~5%, which was the first effective PSC device in our group. Later, they found the same effect of compact TiO$_2$ hole-blocking layer in SnO$_2$-based PSCs where the ETL was composed of mesoporous SnO$_2$ single crystal. A relatively low PCE (3.8%) was obtained due to the strong charge recombination at the SnO$_2$/perovskite interface. By coating a thin TiO$_2$ hole-blocking layer on SnO$_2$ and TCO substrate via TiCl$_4$ treatment, the PCE was increased to 8.5%. The thin TiO$_2$ hole-blocking layer was demonstrated to considerably reduce the recombination while largely maintaining the superior electron transport properties of the mesoporous SnO$_2$ nanocrystals, thus improving the PCE efficiently. Nowadays, this TiO$_2$ compact layer is not necessary for the SnO$_2$-based or other planar n-i-p structure devices because of the advancement in the film quality of ETL and perovskite. However, better physicochemical contact at the TCO substrate interface is still necessary for high-efficiency electron transport and high-performance devices. For example, Snaith et al. employed graphene as an interlayer between FTO substrate and TiO$_2$ ETL, which exhibited beneficial effect on electron transport because the band position of graphene was suited for electron cascade. Another method to engineer the TCO/ETL interface is to modify the work function (WF) of TCO. Zhou et al. demonstrated that surface modification with an amine-containing polyethylenimine ethoxylated (PEIE) could give rise to the formation of an interfacial dipole with preferred direction caused by the slight electron transfer from the amine-containing molecules to the TCO surface, which contributed to the reduction in the WF of TCO substrate and thus the better electron transport. This way of engineering the interface by introducing an interlayer and optimizing the interface energy structure is an universal strategy for various kinds of interfaces in PSCs.

3.3.3 Interface between ETL and Perovskite

TiO_2 is a typical ETL material for the conventional n-i-p structure PSCs, which performs the function of electron extraction and transport. It is found that the low carrier mobility and the generation of deep traps by UV light in TiO_2 could induce charge accumulation, recombination and thus the IV hysteresis, damaging the device performance. It is necessary to control the electron extraction, transport, and recombination at the interface between the TiO_2 ETL and perovskite layers. Doping TiO_2 ETL with metal aliovalent cations, such as Li, Mg, Y, Al, and Nb, was an efficient way to improve the properties of both TiO_2 ETL and TiO_2/perovskite interface, while introduction of interlayer achieved huge success as well. Snaith et al. first employed a C60 monolayer onto the mesostructured TiO_2 layer to modify the TiO_2 ETL/perovskite interface, which could enhance charge separation, reduce the capacitance of TiO_2, thus diminishing the IV hysteresis. Meng et al. constructed a TiO_2/graphdiyne/perovskite interface where graphdiyne can effectively passivate grain boundaries and interfaces, thus facilitating the photogenerated electron extraction. Yang et al. spin coated a triblock fullerene derivative [6,6]-phenylC61-butyric acid-dioctyl-3,3'-(5-hydroxy-1,3-phenylene)-bis (2-cyanoacrylate) ester (PCBB-2CN-2 C8) onto the TiO_2 ETL surface to reduce the trap states of TiO_2. They demonstrated that the oxygen vacancies of TiO_2 responsible for hysteresis can be passivated by the adsorption of electron-withdrawing groups (−C≡N and carbon balls). Because of the introduction of PCBB-2CN-2C8, the WF of TiO_2/PCBB2CN-2 C8 was reduced to 4.01 eV from 4.21 eV of TiO_2, giving rise to the improvement in the open-circuit (V_{oc}), FF, and finally the PCE. Our successful approach was to introduce an ultrathin graphene quantum dots (GQDs) layer between perovskite and TiO_2 (Fig. 3.4). These GQDs were single- or few-layer graphene but possessed a tiny size of only several nanometers with special quantum confinement and edge effects as compared to both conventional quantum dots and graphene (Fig. 3.4d). Furthermore, the CBM position of our GQD was 0.2 eV higher than the TiO_2 ETL (Fig. 3.4b). This modification in the surface energy structure was beneficial for the electron extraction between perovskite and TiO$_2$ ETL. Ultrafast transient absorption spectroscopy revealed that there was a considerably faster electron extraction time (90–106 ps) with GQDs-based perovskite/TiO$_2$ film, as compared to 260–307 ps of pristine perovskite/TiO$_2$ film (Fig. 3.4e). As a result, the PSC based on this GQD interface layer exhibited a PCE of 10.2%, higher than that of the device without inserting GQDs (8.8 %). In addition to inserting a modifying layer between ETL and perovskite layers, researchers also introduced a hierarchical dual scaffold consisting of a quasi-mesoscopic inorganic TiO$_2$ layer and a percolating organic PCBM manifold throughout the capped or filled perovskite bulk to improve the photogenerated charge separation and collection efficiencies (Fig. 3.5). It was found that the soft PCBM scaffold exhibited efficient charge separation due to the formation of an interpenetrating network intimately interfaced with perovskite crystals, meanwhile the quasi-mesoporous hard TiO$_2$ scaffold provided a continuous pathway for electron transport (Fig. 3.5c,f). It was because of this optimum electron transport

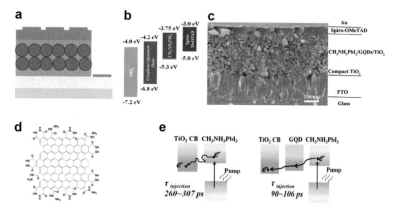

Figure 3.4 Schematic structure of the GQDs interlayer-based PSC (a), where the mesoporous oxide is either loaded with GQDs, the energy band alignment relative to vacuum (b), cross-sectional SEM image of a complete photovoltaic device (c), the edge-modified GQD structure determined by theoretical calculation (d), and schematic illustration of electron generation and extraction at TiO$_2$/MAPbI$_3$ and TiO$_2$/GQDs/MAPbI$_3$ interfaces (e).

paths orthogonal to the hole transport paths (Fig. 3.5f), that the semitransparent PSCs based on an ultrathin perovskite layer (only about 100 nm) with this dual-scaffolds could achieve an internal quantum efficiency of ~100% and a PCE of 12.3%, placing among the highest performing devices of the kind reported at that time.

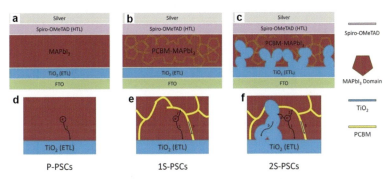

Figure 3.5 Schematic illustration of (a) structure and (d) charge separation in conventional planar p-i-n PSCs (P-PSCs), (b) structure and (e) charge separation in PSCs with PCBM scaffold (1 S-PSCs), (c) structure and (f) charge separation in PSCs with organic (PCBM)-inorganic (TiO$_2$) scaffolds (2 S-PSCs). (Qiu and Yang, 2019).

The modification or post-treatment of TiO$_2$ ELT surface through introducing a fullerene- or graphene-based interlayer can also be applied to the SnO$_2$ ETL-based PSCs. Caruso et al. constructed an SnO$_2$/C60/perovskite interface where the back reaction between the injected electrons in ETL and the holes in the perovskite layer can be effectively restrained, suppressing the charge recombination and enhancing the device performance. Jen and Yan assembled a C60 monolayer onto the SnO$_2$ ETL surface, demonstrating that the C60 monolayer can significantly passivate the surface defects of SnO$_2$, thus enhancing the electron transport efficiency. Fang et al. synthesized an SnO$_2$/PCBM bilayer ETL, of which the PCBM can passivate the grain boundaries and surface of perovskite, obtaining a PSC with high PCE value and reduced hysteretic IV behavior. Like what we did in TiO$_2$ ETL-based PSCs, Yu et al. also applied the GQD as the interlayer between SnO$_2$ ETL and perovskite. They demonstrated that the photogenerated electrons in GQD can transport to the CB of SnO$_2$, which can fill the electron trap states, improving the

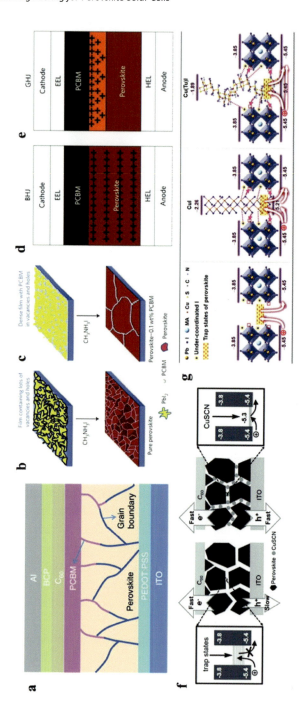

Figure 3.6 (a) Device structure with PCBM migrated into perovskite grain boundaries. (b,c) The mechanism for the formation of perovskite grains in the absence and presence of PCBM. The schematic of (d) bulk heterojunction and (e) graded heterojunction. (f) Schematic illustration of working mechanisms of the devices without or with CuSCN incorporated in perovskite layer. (g) Schematic illustration of possible mechanism for the trap state passivation. (h) The formation of quasi-3D FAxPEA1−xPbI3 crystal. (i) The proposed thermal degradation routes of the different devices with or without 3D–2D graded interface. (Qiu and Yang, 2019).

Fermi level and conductivity of SnO_2. The electron extraction efficiency was eventually enhanced, giving rise to the suppression of the recombination at the ETL/perovskite interface. Besides the fullerene- or graphene-based materials, He et al. spin coated a triphenylphosphine oxide (TPPO) layer onto the SnO_2 surface to reduce the WF of SnO_2, which can improve the built-in field, thus decreasing the energy barrier at the SnO_2/perovskite interface and finally enhancing the photogenerated electron transport efficiency (Zheng et al., 2019).

3.3.4 Grain Boundaries in the Perovskite Active Layer

A perovskite film often consists of perovskite nano- or macrocrystals separated by grain boundaries. The grain boundaries of the perovskite film were always recognized as recombination sites leading to the decrease in device performance and hysteresis phenomenon, so the modification of interface between grain boundaries was also widely studied (Fig. 3.6). In 2014, Huang et al. treated the device with post-thermal annealing to introduce the PCBM molecule into the grain boundaries of perovskite film; the passivation effect reduced the interface charge recombination and minimized the photocurrent hysteresis in the p-i-n PVSC. Finally, Wu and Chiang directly mixed the PCBM molecule within the perovskite film to form a bulk-heterojunction p-i-n device. The PCBM molecule was incorporated at or near perovskite grain boundaries, making a significant impact on electronic properties and balancing the electron and hole mobility with long charge diffusion length in terms of the highly enhanced fill factor, J_{sc}, and hysteresis. Similar work was also demonstrated by Sargent et al. on regular structure, which was further applied into semitransparent devices. Han et al. introduced the PCBM into perovskite layer through the solvent-dropping process, which helped to form a graded heterojunction structure. This structure improved the photoelectron collection and reduced recombination loss, resulting in a certified efficiency exceeding 18% on p-i-n PVSC with aperture area greater than 1 cm^2. Besides the ETMs, HTMs were also used to passivate the perovskite grain boundaries. Huang et al. deposited the perovskite with HTM CuSCN together and found that the CuSCN incorporated in the perovskite grain boundaries, which build a pathway for hole transportation, effectively facilitated the hole transfer from perovskite to ITO electrode. Finally, they used Cu(thiourea)I as trap state passivator to interact with the undercoordinated metal cations and halide anions at perovskite crystal surface and grain boundaries. Owing to the matched valancing band maximum, the trap state passivator/perovskite-formed bulk heterojunction (BHJ) facilitated the hole transport and decreased the charge recombination. Jen et al. introduced a phenylethylammonium iodide (PEAI) into perovskite, where the PEA$^+$ can passivate the defects at lattice surface and grain boundaries and behaving as molecular locks to strengthen the intermolecular interactions of perovskite. In addition, perovskite also suffered from ion migration, and recent studies showed that grain boundaries might be the main sites for this process. In the work of Wang et al., they clearly demonstrated the reconstruction of grain boundaries caused by ion migration, which may induce defects in both perovskite film and perovskite/PCBM interface. Recently, our group developed a graded 3D–2D perovskite interface with 2D structure passivated grain boundaries. We found that the 2D perovskite could passivate at the surface and grain boundaries of perovskite film, which modified the surface energy level and reduced the charge recombination, resulting in an ultrahigh V_{oc} of 1.17 V. And at the same time, the 2D passivated grain boundaries effectively blocked the cross-layer ion diffusion, prevented the migration of I$^-$ ion from perovskite layer to PCBM layer, thus dramatically improved the thermal stability of p-i-n PSC.

3.3.5 Interface between Perovskite and HTL

It was demonstrated that the PSCs constructed with typical HTL of spiro-OMeTAD have excellent charge separation properties at the perovskite/spiro-OMeTAD interface, which can be partially ascribed to the improved hole conductivity in spiro-OMeTAD caused by the doping of cobalt electrolyte and lithium salts. A later study revealed that the high hydroscopicity of the doped lithium salts could induce the decomposition of organometal halide perovskite, thus giving rise to the damage of the interface between perovskite

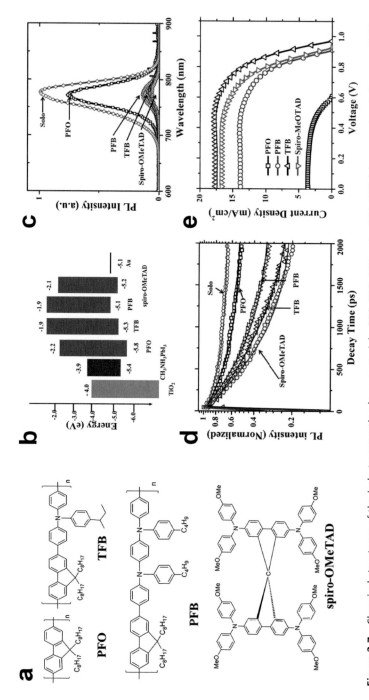

Figure 3.7 Chemical structures of the hole-transporting layer materials: PFO, TFB, PFB, spiro-OMeTAD (a), energy band diagram of TiO$_2$, perovskite, PFO, TFB, PFB, spiro-OMeTAD, and Au electrode (b), steady-state PL spectra (c) and PL decay spectra (d) of PFO, PFB, TFB, spiro-MeOTAD on TiO$_2$/MAPbI$_3$ film and contrast film with solo TiO$_2$/MAPbI$_3$. (e) Current density–voltage curves for the best-performing solar cells with different HTL, PFO (square line), PFB (circle line), TFB (up triangle line), spiro-OMeTAD (down triangle line). (Qiu and Yang, 2019).

and HTL and failure of the corresponding devices. An ideal hole transport material (HTM) should fulfill high hole mobility and the compatible HOMO energy levels to the VBM of organometal halide perovskites to construct an efficient interface for photogenerated charges separation and hole transport. Besides, the stability is another important factor since the HTL plays a role of covering layer on perovskite layer, which is sensitive to the humidity. PTAA, an arylamine-derived conjugated polymer, is a good alternative to spiro-OMeTAD owing to its high hole mobility, related strong physicochemical interaction with perovskite. The conventional n-i-p PSCs based on PTAA have exhibited a high PCE above 20%, which is comparable to the spiro-OMeTAD-based ones. One strategy was to introduce a low-cost polyfluorene derivative polymer containing fluorine and arylamine group (TFB) as HTL in the conventional n-i-p PSCs (Fig. 3.7). The HOMO position of our TFB was 0.1 eV lower than the spiro-OMeTAD (Fig. 3.7b), resulting in an optimized interface energy structure for hole extraction. Photoluminescence (PL) indicated an efficient hole extraction and diffusion features at the interface between perovskite and TFB (Fig. 3.7c,d). As a result, the fill factor (FF), photocurrent, and open-circuit voltage of the TFB-derived cells was higher than those of the ones with spiro-OMeTAD, establishing that TFB is a potential contender for constructing high-efficiency perovskite/HTL interfaces and thus the high-performance PSC devices.

3.4 Interface Engineering of Inverted p-i-n Structure PSCs

3.4.1 Current–Voltage Hysteresis Problem: From n-i-p to p-i-n Structure PSCs

For the conventional n-i-p PSCs constructed by using TiO_2, ZnO as ETL, the devices usually exhibited higher performance under reverse scan (open-circuit to short-circuit scan) than under the forward scan (short-circuit to open-circuit scan). Such hysteresis behavior in IV curves was usually more obvious in planar n-i-p device than the mesostructured n-i-p device, making it difficult to evaluate the real device performance. Charge trapping at the perovskite grain boundaries and interfaces, ion migration, and ferroelectric effects were proposed to be responsible for the IV hysteresis. Generation of deep traps by UV light in TiO_2 that can capture electrons, inducing charge accumulation, was another origin of the IV hysteresis, especially for the TiO_2 ETL-based devices. Nowadays, charge accumulation induced by ion migration in the perovskite layer is considered to be the main reason for the IV hysteresis behavior. However, IV hysteresis is negligible in the inverted p-i-n PSCs based on fullerene ETL due to the passivation effects. It has been demonstrated that fullerenes could passivate the trap states, leading to the formation of a fullerene halide radical, which can block the ion migration. Fullerene molecules could also diffuse into the perovskite layer through pinholes and grain boundaries, increasing the contact area of perovskite/fullerene interface, which can improve charge transfer at the interface, resulting in the quick dissipation of the interface capacitive charges. Besides, fullerenes could extract the electrons more efficiently, thus restraining charge accumulation at the interface and consequently reducing the IV hysteresis. In addition to the passivation effects, fullerene-based inverted p-i-n PSCs demonstrated improved stability as compared to the TiO_2 ETL-based n-i-p PSCs, because the oxygen vacancies in TiO_2 can be activated under UV illumination, leading to the generation of O_2^-, which could decompose the organometal halide perovskites. Finally, the inverted p-i-n structure device can be produced in low temperature range, making it compatible with flexible photovoltaics. These striking properties inspired researchers to develop material and interface for inverted PSCs, which will be discussed in following.

3.4.2 Interface between TCO and HTL

The inverted p-i-n structure PSCs can be regarded as originated from organic polymer bulk-heterojunction solar cells. The first p-i-n structure PSCs were fabricated by sandwiching the perovskite with fullerene derivative ETL and transparent HTL of PEDOT:PSS, which is widely used in organic photovoltaics and LEDs. However, PEDOT:PSS was a less than ideal candidate owing to its acidity, tendency to absorb water, inability to block electrons, and low WF, which could result in an inefficient interface for hole extraction. Modification of PEDOT:PSS with polar solvent additives, including DMSO, DMF, and EG, could give rise to the improvement in the conductivity and morphology of the as-prepared layer, while doping the PEDOT:PSS with additives of low LUMO level could improve both the conductivity and WF, thus reducing the energy barrier for hole extraction at the interface. In addition to PEDOT:PSS, other p-type semiconductors such as CuSCN, Cu_xO, graphene oxide, and PTAA have also been utilized as the HTL in p-i-n PSCs. Earlier work introduced a new-type nickel oxide (NiO) nanocrystalline (NC) as a transparent efficient HTL for inverted p-i-n structure PSCs (Fig. 3.8). The HTL constructed by this NiO NC prepared by simple sol-gel process revealed a rough film surface formed by the aggregation of faceted NiO NCs, which can permit the formation of an intimate, large interfacial-area junction with the perovskite film, thus establishing an efficient interface for hole extraction. Besides, the VBM (−5.36 eV) of the NiO NC HTL was suitable for the hole extraction from the $MAPbI_3$ layer, whose VBM is −5.4 eV as compared to the PEDOT:PSS layer (−5.21 eV) (Fig. 3.8a). This appropriate energy structure at the interface was beneficial for hole extraction and electron blocking. PL analysis demonstrated that the hole extraction and transport capabilities of this NiO film interfaced with the $MAPbI_3$ film were higher than those of traditional organic PEDOT:PSS layers (Fig. 3.8b,c), while the film thickness played a curial role in device performance (Fig. 3.8d). It was found that the device with an NiO NC film with a thickness of 30–40 nm exhibited the best performance, as a thinner layer led to a higher leakage current, whereas a thicker layer resulted in a higher series resistance. Eventually, researchers fabricated an inverted structure device based on this NiO NC HTL with a PCE of 9.1%, which was high for the planar inverted p-i-n PSCs based on inorganic HTL at that time.

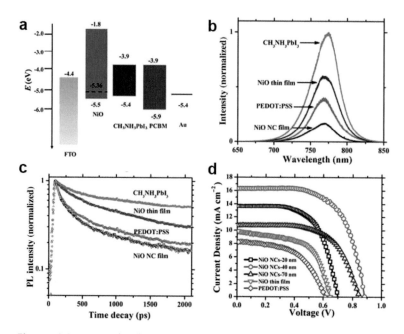

Figure 3.8 Energy-level alignment diagram of the NiO-based device components (relative to the vacuum level) (a); the dashed line shows the Fermi level of NiO NCs from the UPS measurement. Steady-state PL spectra (b) and transient PL decay (c) of NiO NCs (40 nm thickness), PEDOT:PSS, and an NiO thin film interfaced with an MAPbI$_3$ film. The bare MAPbI$_3$ film was used as a reference. Typical J–V curves of the PSCs based on NiO films with different thicknesses (d). (Qiu and Yang, 2019).

3.4.3 Interface between HTL and Perovskite

Compared with PEDOT:PSS HTL, NiO$_x$ have demonstrated advantages of excellent stability and electron-blocking ability. However, the photovoltaic performance of NiO$_x$-based inverted PSCs is limited because of the non-sufficient contact between NiO$_x$ and perovskite and the low carrier transport properties of the NiO$_x$ NCs-based films. A series of dopants have been applied to improve the conductivity of NiO$_x$, such as Cu, Li, Cs, and Co. He et al. demonstrated that molecular doping and post-treatment of alkali chlorides were effective ways to improve the hole transport properties both in the NiO$_x$ HTL and NiO$_x$/perovskite interface. Researchers also developed a series of other methods, including surface modification, hybridization of

NiO$_x$ with metal nanoparticle, and grain boundary engineering of perovskite to optimize the interface properties such as energy-level alignment and electronic contact features. They first introduced diethanolamine (DEA) as an interlayer modifier to improve the interface connection through perovskite-hydroxyl and NiO$_x$-amine groups (Fig. 3.9). The DEA interlayer was prepared by dipping the NiO$_x$ film in a DEA isopropanol (IPA) solution for several minutes, rinsed with IPA, and then heated at 100°C for 10 min in N$_2$-filled glove box. XPS analysis revealed that the WF of NiO$_x$-based HTL was slightly decreased from 4.47 to 4.41 eV after the introduction of DEA monomolecular on NiO$_x$ surface (Fig. 3.9b,c), which should be attributed to the chemical interaction between DEA and NiO$_x$. As the N atom of DEA is linked with the Ni of NiO$_x$ film, the −NH− could participate in chemisorption and lower the WF of NiO$_x$, giving rise to the formation of a favorable molecular dipole layer and thus the optimized interface energy-level alignment for enhancing the hole extraction rate and charge transport rate (Fig. 3.9c), as proved by the transient PL and IS measurements (Fig. 3.9d–f). As a result, the PCE of the PSCs was considerably promoted to ~15%. Besides, the device produced from DEA modification strategy also possessed improved stability and hysteresis-free properties, presumably ascribed to the improved interface contact induced by the DEA interlayer. Recently, they developed an NiO$_x$–Au composite layer to improve the hole transport properties of the NiO$_x$-based HTL through embedding a small concentration (0.11 At%) of gold nanoparticles (Au-NPs), 2–3 nm in diameter, into the NiO$_x$ thin film. The Ohmic contact nature of the Au–NiO$_x$ junction could promote the transport of electrons from NiO$_x$ to Au, in effect resulting in an equivalent increase in the hole concentration in NiO$_x$. The hole concentration in Au-embedded NiO$_x$ was thus tripled compared to the non-Au-embedded one, which in turn induced the corresponding shift in Fermi level and VBM, leading to the formation of a favorable interface energy-level alignment for hole extraction and transport. Time-of-flight secondary ion mass spectrometry (ToF-SIMS) depth profile measurement demonstrated that the concentration of Au in NiO$_x$ thin film was only 0.11 At%, which could avoid the direct contact between Au and perovskite that would aggravate carrier recombination. Eventually, the PSCs constructed with this NiO$_x$–Au HTL revealed a high PCE of ~20%, which was among the best performed p-i-n structure PSCs.

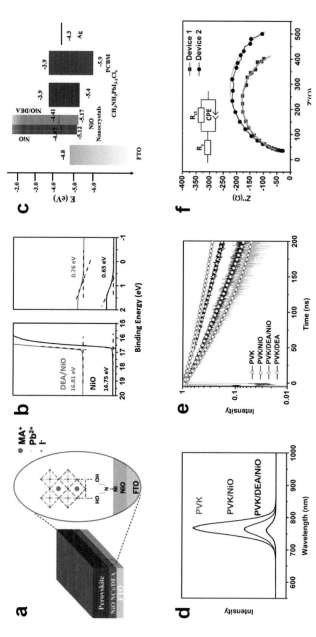

Figure 3.9 (a) Schematic illustration of the surface modification of the NiO$_x$ NC film with a DEA monolayer, highlighting an enhanced contact at NiO$_x$/ perovskite interface through a link between –OH group and Pb in perovskite. (b) UPS of bare NiO$_x$ and NiO$_x$/DEA films. (c) Schematic energy levels of each layer in perovskite solar cell. The Fermi levels of NiO$_x$ and DEA/NiO$_x$ are represented by dashed lines. (d) Steady-state PL spectra (linear plot) and (e) normalized transient PL decay profiles (logarithmic plot) of perovskite layers on the NiO$_x$, NiO$_x$/DEA, and quartz substrates. The transient PL decay for the perovskite layer on quartz/DEA is also included in panel (e). (f) IS Nyquist plots obtained under the short-circuit and full-sun illumination (AM1.5, 100 mW/cm^2) condition. (Qiu and Yang, 2019).

In addition to the construction of suitable NiO_x-based HTL, we also developed a solvent-vapor-assisted post-treatment strategy to improve the perovskite crystal quality for achieving high-efficiency HTL/perovskite interface and photovoltaic devices (Fig. 3.10). Our work found that there was a type of $MA_2Pb_3I_8(DMSO)_2$ intermediate phase formed when the as-prepared $MAPbI_3$ film is annealed in DMSO atmosphere. This $MA_2Pb_3I_8(DMSO)_2$ intermediate phase then decomposed at the $MAPbI_3$ grain boundaries, which facilitated grain boundary migration inside the perovskite film (Fig. 3.10a). As a result, the as-prepared $MAPbI_3$ grains grew to larger sizes under DMSO atmosphere compared to those under N_2 atmosphere (Fig. 3.10b). From Fig. 3.10c, one can see that these large sizes of $MAPbI_3$ grew perpendicularly to the substrate, enabling the improved HTL/perovskite interface contact, which can promote the charge transport of the as-prepared devices. Adding the MABr additive into DMSO vapor can further promote grain boundary migration and at the same time embedded Br into the pristine $MAPbI_3$ film, thus resulting in the formation of a gradient Br-rich perovskite phase in the interface region between the $MAPbI_3$ and NiO_x layers, which may serve to passivate this interface (Fig. 3.10d,e). Eventually, the inverted planar structure device with this Br-doped perovskite produced under DMSO/MABr vapor indicated a considerably higher current density (21.67 mA/cm^2) and open-circuit voltage (1.10 V) than those of N_2-produced one, resulting in an impressive PCE of 17.6%. Researchers also developed another method to optimize the HTL/perovskite interface via modification of the crystal properties of the as-prepared perovskite films. They conducted a systematic and in-depth study on the intermediate film-assisted crystallization of perovskite on planar NiO_x substrates via a strict control over the DMSO/DMF ratio in $MAPbI_3$ solutions. A series of intermediate films with different compositions was produced, including perovskite, perovskite/$MA_2Pb_3I_8(DMSO)_2$, and $MA_2Pb_3I_8(DMSO)_2$, which could afford perovskite crystals via down-growth, down- and up-growth, and up-growth mechanisms, respectively. It was found that the up-growth perovskite crystals ascribed to the pure $MA_2Pb_3I_8(DMSO)_2$ demonstrated the best interface contact with NiO_x HTL, optimal alignment without horizontal grain boundaries, which facilitated charge transfer and reduced charge recombination. The inverted PSCs constructed with this $MA_2Pb_3I_8(DMSO)_2$ intermediate-evolved perovskite revealed a high PCE of 18.4% and improved stability.

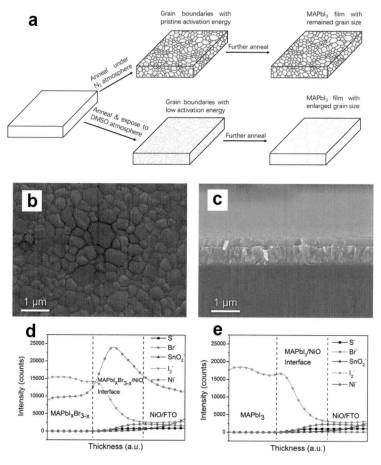

Figure 3.10 (a) Scheme of the proposed morphology evolution of $MAPbI_3$ grains under different atmosphere. SEM images of $MAPbI_xBr_{3-x}$ films annealed under MABr and DMSO mixture atmosphere: (b) top view, (c) side view. The as-prepared perovskite film shows a columnar structure. ToF-SIMS depth profiles of $MAPbI_xBr_{3-x}$ film annealed under (d) MABr and DMSO mixture atmosphere, and (e) MABr-only atmosphere. The film annealed in the mixture atmosphere shows a successful embedment of Br with a relative intensity of Br at 10^4, whereas the film annealed in the MABr-only atmosphere shows no trace of Br element. (Qiu and Yang, 2019).

3.4.4 Interface between Perovskite and ETL

The perovskite/ETL interface is related to electron extraction, hysteresis, and device stability in inverted p-i-n PSCs. PCBM is the widely used ETL in inverted structure devices. However, due to the insufficient viscous solution for spin coating caused by the small molecule of PCBM, it was difficult to form a high-quality PCBM film on the relatively rough perovskite surface in the earlier stage of the inverted structure device fabrication, which thus induced the inefficient interface charge transfer. In this regard, researchers developed a route to produce high-quality PCBM ETL by adding a small percentage (1.5 wt%) of high molecular weight polystyrene (PS) into the PCBM ETL. It was found that the addition of PS promoted the formation of a highly smooth and uniform PCBM ETL via spin coating, which can prevent undesirable electron–hole recombination between the perovskite layer and the top electrode, and thus significantly enhanced the PCEs of corresponding cells. In addition to the PS doping method, we also introduced a new fullerene derivative named C5-NCMA as an ETL to replace the commonly used PCBM in the inverted PSCs. Compared with PCBM, C5-NCMA featured a higher hydrophobicity, higher LUMO energy level, and higher ability of self-assembly for film processing, which favored the formation of efficient interface for electron transfer and transport. The PSC with the device structure of FTO/NiO$_x$/MAPbI$_3$/C5-NCMA/Ag showed a PCE of up to 17.6% with negligible hysteresis, which is higher than PCBM-based one (16.1%). Importantly, the stability of the C5-NCMA ETL-based device to moisture was significantly enhanced compared to the PCBM-based one due to the hydrophobic nature of C5-NCMA.

Modifying the perovskite layer properties is another way to optimize the properties of the interface between perovskite and ETL, since the defects and the grain boundary properties at the side of perovskite can affect the interface properties significantly. Another issue one should pay attention to is the ion migration in perovskite, which takes place mainly in the grain boundaries. Wang et al. demonstrated that the reconstruction of grain boundaries caused by ion migration could induce defects in both perovskite film and perovskite/PCBM interface. Jen et al.

introduced phenylethylammonium iodide (PEAI) into perovskite, where the PEA$^+$ can passivate the defects at lattice surface and grain boundaries and behaving as molecular locks to strengthen the intermolecular interactions of perovskite. Meanwhile, Yang et al. developed a hybrid perovskite with a graded 3D–2D perovskite interface, where the 2D structure perovskite was introduced to passivate grain boundaries of 3D perovskite (Fig. 3.11). This hybrid graded 3D–2D perovskite layer was prepared by replacing the pure toluene by PEAI/toluene solution in the conventional solvent dripping process prior to conversion to a crystallized perovskite film by annealing at 95°C (Fig. 3.11a). SEM analysis demonstrated that the PEAI-assisted toluene dripping process did not seem to affect the quality of the crystalline films. ToF-SIMS depth profiles for different perovskite films revealed that the PEAI was introduced into the MAPbI$_3$ layer during the solvent dripping process (Fig. 3.11e–g). Remarkably, the intensity of PEA$^+$ fragment decreased steeply at near surface of the film and then settled at a steady level much lower than MA$^+$, indicating the successful deposition of an ultrathin layer of PEA$^+$ on the perovskite surface (Fig. 3.11f). The morphology and composition distribution for each kind of perovskite films are shown in Fig. 3.11h–j. The 3D–2D graded perovskite film (Fig. 3.11i) was mainly composed of high-quality MAPbI$_3$ crystals, but what made it unique was the formation of an ultrathin 2D perovskite layer in the near-surface region. The introduction of 2D perovskite could modify the interface energy level, thus reducing the charge recombination and increasing the PCE values of the corresponding devices (Fig. 3.11k–m). Furthermore, the 2D perovskite could block the cross-layer ion diffusion effectively, which in turn prevented the migration of iodide ions from the perovskite layer to the PCBM layer, thus improving the thermal stability dramatically.

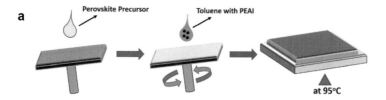

Interface Engineering of Inverted p-i-n Structure PSCs | 77

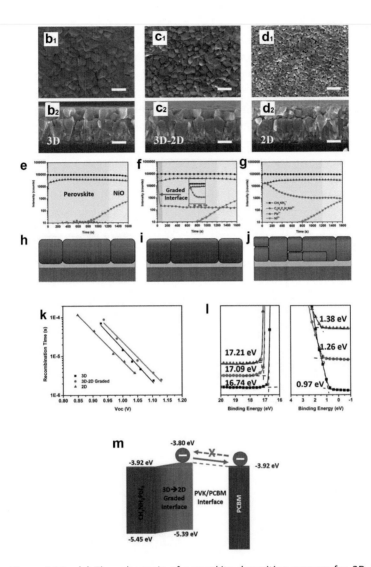

Figure 3.11 (a) The schematic of perovskite deposition process for 3D–2D graded perovskite film. The top-view and cross-sectional SEM images of (b) 3D, (c) 3D–2D graded, and (d) 2D perovskite films. The scale bar is 500 nm. ToF-SIMS depth profiles and schematics of (e, h) 3D, (f, i) 3D–2D graded, and (g, j) 2D perovskite films on NiO/FTO substrates. (k) IMVS derived carrier recombination lifetimes for the different devices. (l) UPS (He I) of the perovskite sample films. (From top to bottom: 2D, 3D–2D Graded, and 3D sample. (m) Schematic of energy-level alignment at the perovskite and PCBM interface. (Qiu and Yang, 2019).

3.4.5 Interface between ETL and Metal Electrode

For the inverted p-i-n structure PSCs, an interface barrier usually exists between the ETL (e.g., PCBM) and metal electrode (e.g., Ag, Al), which results in the poor electron extraction efficiency. Organic molecules, such as bathocuproine (BCP), 4,7-diphenyl-1,10-phenanthroline (Bphen), perylene diimide (PDINO), and metal acetylacetonates are typically introduced as the interlayer between fullerene derivatives and metal electrodes to improve the electron extraction efficiency. Besides the organic interlayer, inorganic materials, such as SnO_2 and Al-doped ZnO, can also be utilized as interface materials and were proved to work well. It was found that the FF commonly increased with the introduction of these electrode interfacial layers mentioned earlier. Meanwhile, Yang et al. reported the synthesis of a tailor-made, amino-functionalized perylene diimide polymer (PPDIN6) and its demonstration as a new interlayer material between PCBM and metal electrode in inverted p-i-n PSCs. Due to the reduced trap density at the PCBM/Ag interface as elucidated by temperature-dependent admittance spectroscopy measurement, the charge recombination was suppressed, thus giving rise to the enhancement of electron extraction and the photoelectric conversion. Besides, the amine groups in PPDIN6 can neutralize the migrating iodide ions and inhibit the formation of the insulating Ag–I bonds on the surface of the Ag electrode, leading to an improved device stability as compared to the device without the PPDIN6 interlayer.

3.5 Interface Engineering of C-PSCs

3.5.1 Brief Introduction of C-PSCs

Carbon electrode-based PSCs (C-PSCs) were initially developed from traditional n-i-p structure PSCs. Here the organic HTL and metal electrode are replaced by the carbon-based materials, which play the role of hole extraction and transport. The first C-PSC was constructed by Miyasaka et al., with a low PCE value of 0.37%. Han et al. largely improved the device performance to PCE of ~10% by developing a mesostructured cell in which a TiO_2 scaffold layer, a porous ZrO$_2$ insulating layer, and a porous carbon electrode were sequentially deposited, followed by infiltration in a perovskite precursor solution. Nowadays, C-PSCs have achieved PCE of above ~15% and exhibited promising advantages for commercialization, such as inert to ion migration originating from perovskite and metal electrodes, inherently water resistant, easy fabrication, etc. The device structure and working principle of C-PSCs are different from the metal electrode-based ones, and some specific and innovative strategies must be employed to improve the perovskite layer and interface properties. In the following, we will summarize a series of work on developing efficient materials and interfaces for high-performance C-PSCs.

3.5.2 Interface between Oxide Semiconductor ETL and Perovskite

In traditional metal electrode-based PSCs, solar light passing through the perovskite layer can be reflected by a smooth metal electrode, enabling a secondary absorption. Consequently, the thickness of perovskite layer is usually in the relatively low range (400–600 nm). However, black carbon electrodes cannot reflect passing light; a thicker perovskite layer (~1 μm) is thus typically introduced to complete light absorption in C-PSCs. As a result, the TiO$_2$ ETL needs to be made porous with larger thicknesses (400–600 nm) than those in metal electrode-based PSCs (100–200 nm). A thick porous TiO$_2$ film means increasing in the interface area and strong carrier recombination, which would restrain the perovskite pore-filling and slow down charge collection. To balance the inefficient charge transport in thick, porous TiO$_2$ film and the inefficient light harvesting in thin perovskite film, Yang et al. designed a TiO$_2$ nanobowl (NB) array film as an alternative ETL for C-PSCs by combining the colloidal template and liquid phase deposition methods. The TiO$_2$ NB array films with adjustable ordered structure and thickness were fabricated by changing the polystyrene (PS) diameter in the monolayer packing of PS spheres and the sol-gel processing conditions, which can promote the light-harvesting efficiency and gave a direct way for the photogenerated electrons. Besides, the ordered macroporous structure with hollow NB allowed perovskite to easily penetrate the ETL and thus formed an intimate perovskite/TiO$_2$ interface for rapid electron extraction and transport. Leveraging these advantages, the 220 nm PS templated TiO$_2$ NB-based devices performed the best on both light absorption capability and charge extraction, and achieved a PCE up to ~12% with good stability, which was 37% higher than that of the planar counterpart.

TiO$_2$ is the most widely used ETL in C-PSCs. However, TiO$_2$-based PSCs suffer from instability under light illumination caused by the photocatalytic activity of TiO$_2$ as mentioned earlier and the adventitious presence of water, oxygen, etc., adsorbed onto the TiO$_2$ substrate, which would dampen the commercial application of TiO$_2$-based PSCs. Furthermore, most of the reported high-performance hole transport material (HTM)-free C-PSCs suffer from severe constraint on the open-circuit voltage as compared to the HTM-based ones. Since the open-circuit voltage depends on the difference between Fermi levels of the ETL and the carbon electrodes, the relatively low quasi-Fermi level of TiO$_2$, which can induce inefficient interface properties, is another drawback for the C-PSCs. In this regard, we introduced an ultrathin ferroelectric oxide PbTiO$_3$ layer between the TiO$_2$ ETL and perovskite layer to construct an efficient interface for charge separation and transport (Fig. 3.12). It was found that the ferroelectric PbTiO$_3$ interlayer in the TiO$_2$/perovskite interface could unambiguously introduce a larger internal electrical field provided by the permanent electrical polarization, enhancing the built-in potential (Fig. 3.12b), which eventually suppressed the non-radiative recombination, giving rise to the high-efficiency photogenerated charge transfer (Fig. 3.12e,f), and thus the improved photovoltaic performance. Another strategy we adopted was to introduce a C60 ETL to replace TiO$_2$, forming a so-called all-carbon-based PSC with a device structure of FTO/C60/MAPbI$_3$/carbon electrode. The C60 ETL was found to effectively improve electron extraction, suppress charge recombination, and reduce the sub-bandgap states at the interface with MAPbI$_3$, leading to the improved device performance. Besides, this all-carbon-based PSCs were shown to resist moisture and ion migration, resulting in a much higher operational stability under ambient, humid, and light-soaking conditions.

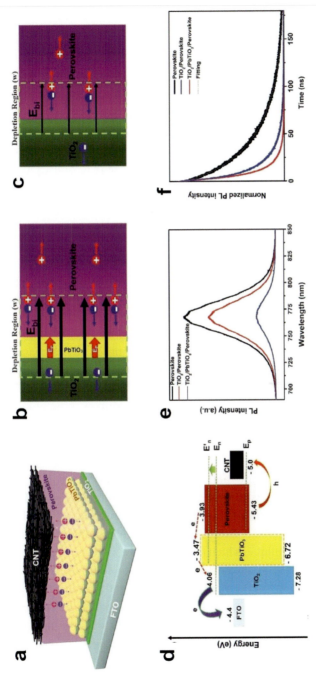

Figure 3.12 (a) Schematic structures of the PbTiO$_3$-interlayer-based C-PSC. Schematic diagrams showing the depletion regions in C-PSCs with (b) and without (c) the PbTiO$_3$ interlayer. (d) Energy-level alignment of the various layers in C-PSCs. (e) Steady-state PL spectra of perovskite layers on TiO$_2$, TiO$_2$/PbTiO$_3$, and quartz substrates. (f) Time-resolved PL decay spectra of perovskite films coated on different substrates. (Qiu and Yang, 2019).

3.5.3 Interface between Perovskite and Carbon-Based Electrode

The original effective C-PSC developed by Han et al. typically comprises an initially deposited multilayer mesoporous structure, which is subsequently infiltrated by a perovskite solution. The carbon electrode for this mesostructured C-PSC is deposited by doctor blade or screen printing techniques using a carbon paste containing graphite, carbon black and a ZrO_2 (or Al_2O_3) nanoparticle binder, followed by annealed at high temperature (e.g., 400°C). Due to the poor contact at the interface between perovskite and carbon electrode, hole extraction is suppressed significantly, which results in a relatively low device performance. To solve this issue, Yang et al. developed an embedment C-PSCs, in which a PbI_2 layer with a mesoporous carbon electrode was first coated before the conversion to perovskite (Fig. 3.13a–c). Then, the carbon-coated PbI_2 layer was immersed into MAI solution to form $MAPbI_3$, which could embed the carbon electrode because of volume expansion after conversion (Fig. 3.13a–c).

The improved interface contact not only significantly increased the device current density but also boosted the FF, ascribed to more efficient hole extraction, all of which finally resulted in an improved PCE. Later, we further improved the weak adhesion of carbon black on PbI_2 in embedment C-PSCs by introducing a novel conversion solution containing carbon black and MAI, which removed the pre-deposition process of carbon electrode (Fig. 3.13d). The deposition of the carbon electrode and chemical conversion of $MAPbI_3$ happened simultaneously by dropwise deposition of the conversion solution onto the PbI_2 layer. Furthermore, this dropping process for the preceding conversion solution could also be implemented by an inkjet printing technique, which made the process more controllable and precise. Yang et al. thus developed a paintable C-PSC, in which an interpenetrating interface between $MAPbI_3$ and carbon electrode was formed, giving rise to a significantly improved mechanical properties of the carbon electrode and an efficient interface for hole extraction. As a result, the device demonstrated a PCE value of ~12%.

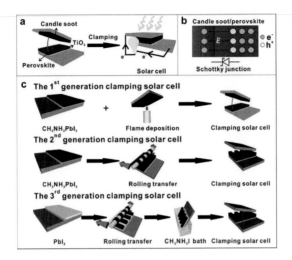

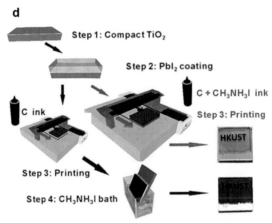

Figure 3.13 Development of embedment clamping C-PSCs. (a) Conception of the clamping solar cells based on a candle soot electrode and a perovskite electrode. (b) The putative Schottky junction formed between candle soot and perovskite, which is at the core of realizing the clamping solar cells. (c) (Top) Fabrication process of the first-generation clamping solar cells by simply clamping an FTO supported candle soot film and an MAPbI$_3$ photoanode. (Middle) Fabrication of the second-generation clamping solar cells by rolling transfer assisted clamping. (Bottom) Fabrication of the third-generation clamping solar cells by chemically promoted rolling transfer clamping, with an MAI bath for the in situ conversion of PbI$_2$ to MAPbI$_3$ partially embedding the soot electrode. (d) Fabrication process of the inkjet printing C-PSCs. For comparison, a different strategy was used to convert PbI$_2$ into MAPbI$_3$ using a separate step 3 and step 4. (Qiu and Yang, 2019).

In the embedment and paintable C-PSCs, the perovskite layer is produced by a two-step spin-coating method, in which the pre-deposited PbI$_2$ is converted to perovskite by dip coating an MAI IPA solution. Unfortunately, the full conversion of PbI$_2$ to perovskite in the conventional MAI IPA solution usually takes a very long time, especially when a thick TiO$_2$ scaffold is used. In consequence, very large perovskite crystals tend to form on the surface due to the Ostwald ripening process, which would result in a poor contact at perovskite/carbon interface and thus the inefficient hole extraction as compared to the HTM-based PSCs. To tackle this problem, Yang et al. developed a novel solvent engineering strategy based on the two-step method to fabricate pure MAPbI$_3$ with an even surface. The cyclohexane (CYHEX) was added to the MAI IPA solution to lower the solution polarity, which not only accelerated the chemical conversion of PbI$_2$ to MAPbI$_3$ to generate a pure perovskite layer, but also suppressed the Ostwald ripening process to achieve an even surface with a compact capping layer. As a result, this even perovskite layer showed enhanced interface contact with the carbon electrode, which successfully improved the device performance and reproducibility. The highest PCE of 14.4% was achieved for the small area device, while the large area device (1 cm^2) achieved a PCE of almost 10%. By applying this solvent engineering strategy to MAPbBr$_3$, we also obtained an even perovskite layer with conformable oriented crystallization, which resulted in an MAPbBr$_3$-based paintable C-PSCs with a record PCE of ~8.1% (V_{oc} = 1.35 V). These studies confirm that it is important to enhance interface contact and improve the PCE of paintable C-PSCs by fabricating a high-quality perovskite layer with even surface.

3.5.4 New-Type Carbon Electrode Materials

For the interface contact or junction between perovskite and carbon electrode, the electric properties at the interface are critical for the hole extraction. It was proposed that there is an Ohmic contact formed at the MAPbI$_3$/Au interface, which did not help to enhance charge separation, giving rise to a low V_{oc} (0.6~0.7 V) for the cell with structure of TiO$_2$/MAPbI$_3$/Au (Au-PSC). However, the replacement of Au with carbon electrode resulted in an improved V_{oc} value (0.9~1.1 V), meaning that the electric properties of the junction formed at the interface between perovskite and carbon are somehow different from those of the perovskite/Au interface. In the work of embedment C-PSCs, we proposed the formation of a Schottky junction at the MAPbI$_3$/carbon interface, as the FTO/carbon/MAPbI$_3$/FTO device exhibited rectifying characteristics (Fig. 3.13b). The built-in field in this Schottky junction helped to increase the overall built-in potential to 0.85 V and presented a direction from MAPbI$_3$ to carbon electrode, which assisted in separating the photogenerated electron–hole pairs so that the holes and electrons could drift toward the carbon electrode and the TiO$_2$/MAPbI$_3$ interface, respectively. The additional built-in field at the MAPbI$_3$/carbon interface explained the higher V_{oc} and PCE achieved by C-PSCs than Au-PSCs. However, it should be noted that whether or not the Schottky junction was formed at all depended greatly on the Fermi levels of perovskite and carbon materials. It was later shown that the electric junction properties at the interface between carbon and perovskite can be adjusted by employing carbon electrodes with different properties. Our work demonstrated that no obvious Schottky junction was formed at the MAPbI$_3$/single-layer graphene (SG) interface due to the similarity between the Fermi levels of MAPbI$_3$ and SG. As a result, the PSCs based on SG achieved an obviously lower performance than those based on multilayer graphene (MG) that could form a Schottky junction with MAPbI$_3$. Therefore, choosing suitable carbon materials is an important way to form an additional Schottky junction at the perovskite/carbon interface and hence to improve device performance. By comparing the embedment C-PSCs with three different carbon materials (carbon black, graphite, and multi-wall carbon nanotubes (MWCNTs)), a clear conclusion was reached that a large difference among these three devices occurred on FF values, with the order of graphite > carbon black > MWCNTs, which led to the same order for the device performance. We further investigated the influence of MWCNT properties on the performance of MWCNTs-based PSCs (Fig. 3.14a–c). Yang et al. showed that boron (B) doping of MWCNT (B-MWCNTs) electrodes was superior in enabling enhanced hole extraction and transport by increasing WF, carrier concentration, and conductivity of MWCNTs (Fig. 3.14c). The C-PSCs constructed by B-MWCNTs as the counter electrodes achieved remarkably higher performances than that with the undoped MWCNTs, with the resulting PCE being improved from

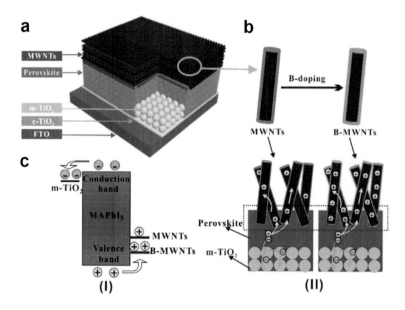

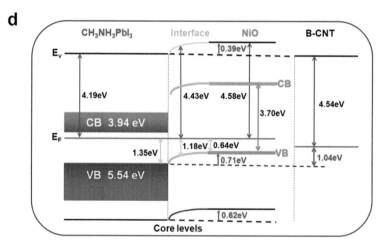

Figure 3.14 Schematic illustration of device configuration, B doping of MWCNTs, and charge behavior in C-PSCs. (a) Schematic illustration of the C-PSC configuration. (b) Schematic diagram of B-MWCNTs. (c) Schematic illustration of charge-transfer enhancement by B-MWCNTs through (I) the lowering the Fermi level of MWCNTs and (II) increasing the number of conduction carriers in the B-MWCNT electrode. The intimate interface between perovskite and MWCNTs is marked by black dotted rectangle in (II). Reproduced with permission. Copyright 2017, American Chemical Society. (d) Energy-level diagram of perovskite/NiO$_x$ interface (middle) and perovskite/MWCNT interface (right). (Qiu and Yang, 2019).

10.7% to 14.6%. Later, Yang et al. developed a method to further optimize the interface between perovskite and MWCNT layer by planting NiO nanoparticles (NP) into the surface region of MAPbI$_3$ crystals via a simple ultrasound spray method, which can form a seamless contact at the perovskite/carbon interface. The NiO NPs layer was found to favorably bend the energy levels at the interface for selective hole extraction and reduction in interfacial recombination (Fig. 3.14d), thereby enhancing the overall photovoltaic performance to a PCE as high as 15.8%, which was among the highest C-PSC efficiencies reported at that time. Besides, the combination of ultrasound spray and nanoparticle embedment technologies opened the way to cost-efficient and scalable production of C-PSCs.

3.6 Conclusion and Perspectives

In this chapter, we have provided a glimpse of interface engineering for high-performance PSCs, primarily based on the selected works. Interface engineering for developing efficient and stable planar PSCs, especially the role of different interlayer materials in the PSCs, is highlighted. Efforts dedicated to optimizing the interfaces have so far involved the following aspects: (1) proper energy-level tailoring between adjacent layers to reduce the energy offset; (2) high charge extraction and transport capacity; (3) the passivation of trap states in the perovskite film; (4) interfacial properties to optimize the perovskite film growth; (5) the protection of the perovskite from moisture; and (6) the protection of the metal electrode from the migrating iodide ions.

The interface properties are considered vital to the fabrication of effective photovoltaic devices in the early development of PSCs and continue to play a curial role nowadays in achieving high PCE and high-stability devices. This echoes emphatically the famous saying, the interface is the device, perhaps even more so with the perovskite solar cells since the organometal halide perovskites may be more prone to defect formation. Specifically, we have discussed strategies and methods for the fabrication of high-quality films, synthesis of nanostructured electrodes, introduction of interlayer materials, modification of work function, etc., in efforts to engineer interfaces for conventional n-i-p, inverted p-i-n, and carbon-based PSCs. To develop excellent interfaces, one first needs to improve the film qualities, control the film thickness of ETL, perovskite and HTL, respectively, thus enhancing the photogenerated electron and hole transport efficiency in each layer of PSCs. Second, intimate contact between each layer is of paramount importance for avoiding large charge recombination at the interfaces. Third, improving the physicochemical contact properties at the interface by constructing interfaces with proper energy-level match between the adjacent layers is indispensable for efficient charge transport with low energy loss, and thus for obtaining high-performance photovoltaic properties. Finally, the interface properties should be engineered in a way commensurate to the device structural features. In this chapter, we have mainly focused on the charge transport properties at the various interfaces of PSCs. Other issues, however, such as, ion migration, moisture invasion, etc., which are curial to the stability of PSC devices, can also be effectively tackled via interface engineering, but need more systematic studies in the future. With a deeper understanding of the interfaces and the development of novel relevant materials and approaches, interface engineering is expected to drive further advance toward ever more efficient and stable PSCs. It is our hope that this chapter will help to facilitate the understanding and the design of interfaces with optimized properties, guiding the future exploration for the eventual commercialization of high-performance PSCs.

Questions

1. Can you describe the relationship between surface structure and some properties of materials?
2. Please list some methods for characterizing surface or interfaces of a material.
3. What is the general principle that determines the electronic structure of a solid–solid interface?
4. How to regulate the properties of material interfaces?

References

Bai Y., Meng X., and Yang S. "Interface engineering for highly efficient and stable planar p-i-n perovskite solar cells." *Adv. Energy Mater.*, **8**(5), 1701883 (2018).

Chen H. and Yang S. "Methods and strategies for achieving high-performance carbon-based perovskite solar cells without hole transport materials." *J. Mater. Chem. A*, **7**(26), 15476–15490 (2019).

Qiu J. and Yang S. "Material and interface engineering for high-performance perovskite solar cells: a personal journey and perspective." *The Chemical Record*, **19**, 1–22 (2019).

Zhang T., Hu C., and Yang S. "Ion migration: a "double-edged" sword for halide-perovskite-based electronic devices." *Small Methods*, 1900552(1-20) (2019).

Zheng S., Wang G., Liu T., Lou L., Xiao S., and Yang S. "Materials and structures for the electron transport layer of efficient and stable perovskite solar cells." *Sci. China Chem.*, **62**(7), 800–809 (2019).

Chapter 4

Carbon Quantum Dot Luminescent Materials

Luminescence is a process by which the energy absorbed by an object in some way is converted into light radiation. The advance of luminescent materials with application in information, energy, materials, aerospace, life sciences, and environmental science and technology will certainly propel the clean energy industry, thus promoting the global and national economy and technology. Semiconductor light-emitting diodes (LEDs) and semiconductor lasers are the two most typical semiconductor luminescence devices. Semiconductor LEDs were born in 1927 and were invented independently by former Soviet scientist Oleg Losev. Rubin Braunstein of the United States Radio Corporation (RCA) observed infrared radiation from diode structures of GaAs and other semiconductor materials at a low temperature of 77 K in 1955. Until 1962, Nick Holonyak Jr. developed the world's first practical red LED. And it was this year that four U.S. laboratories announced almost simultaneously the successful development of a GaAs homogeneous junction semiconductor laser. After that, both semiconductor luminescent diodes and semiconductor lasers have developed extremely rapidly and are quickly widely used in production and daily life. At present, semiconductor luminescent devices have been widely used in the fields of information display, optical fiber

---

Materials and Interfaces for Clean Energy
Shihe Yang and Yongfu Qiu
Copyright © 2022 Jenny Stanford Publishing Pte. Ltd.
ISBN 978-981-4877-66-4 (Hardcover), 978-1-003-14223-2 (eBook)
www.jennystanford.com

communication, solid-state lighting, computer and national defense, and have formed a huge industrial scale.

Semiconductor luminescent materials determine the basic properties of semiconductor luminescent devices. The traditional materials such as Si, Ge, GaAs, and InSb have matured but currently more active is the field of semiconductor luminescent devices. Although they have been developed for hundreds of years, semiconductor luminescent materials and devices are becoming increasingly important in scientific and engineering research. For this reason, this chapter will introduce the semiconductor and semiconductor luminescence basis and take carbon quantum dots (CQDs) as an example to illustrate the latest progress of luminescent nanomaterials.

4.1 Introduction

4.1.1 Semiconductor Physics Basics

This section briefly describes the basic physical properties of semiconductors and introduces some basic concepts of semiconductors, including energy band, direct bandgap, indirect bandgap, p–n junction, and so on, which is the basis of understanding the basic optical properties of semiconductor materials and the working principle of semiconductor luminescent devices.

4.1.1.1 Energy band

In semiconductors, the energy band structure of electrons determines the range of energy allowed and prohibited by electrons and determines the electrical and optical properties of semiconductor materials. The band of the semiconductor can be expressed as in Fig. 4.1. At absolute zero, the highest band that can be filled with electrons forms a valence band. In the valence band, electrons are still bound by various atoms. And above the valence band, electrons can get rid of the binding of a single atom and move the band freely in the whole semiconductor material, that is, the conduction band. For semiconductors, the valence band and conduction band are separated by the forbidden band E_g for bandgap and has $E_g = E_c - E_v$.

The E_c of the formula is the bottom of the conduction band, and the E_v is the top of the valence band.

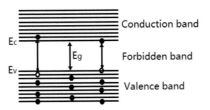

Figure 4.1 Schematic diagram of band structure of semiconductor.

4.1.1.2 Intrinsic and extrinsic semiconductors

Intrinsic semiconductors are ideal semiconductor materials that are pure without any impurities. Because of the thermal vibration of the atoms in the crystal, some electrons in the valence band are excited to the conduction band, while leaving holes in the valence band to form an electron–hole pair. Therefore, the electron concentration n in the intrinsic semiconductor is equal to the hole concentration p.

The introduction of a certain amount of impurities in the intrinsic semiconductor can effectively change the conductive properties of the semiconductor. This semiconductor with a certain amount of impurities is called an extrinsic semiconductor. The introduction of impurity atoms changes the concentration of electrons and holes under thermal equilibrium conditions. However, the concentration of one carrier increases and the concentration of the other carrier decreases. Both intrinsic and extrinsic semiconductors satisfy the law of concentration action under the condition of thermal equilibrium, that is,

$$pn = n_i^2 = N_c N_v \exp(-\frac{E_g}{k_B T}) \tag{4.1}$$

Here, n_i is the intrinsic carrier concentration; $N_c = 2[2\pi m_e^* k_B T / h^2]^{3/2}$ and $N_v = 2[2\pi m_h^* k_B T / h^2]^{3/2}$ are the effective mass of electrons and holes, respectively; k_B is the Boltzmann constant; T is the temperature; and h is the Planck constant.

The formula (4.1) shows that $p = n = n_i$ for intrinsic semiconductors. However, for extrinsic semiconductors due to the introduction of impurities, either the hole concentration is higher than the electron concentration or the electron concentration is higher than the hole concentration. The carriers with high concentrations are called majority carriers, and the carriers with low concentrations are called minority carriers. If majority carriers are electrons, such extrinsic semiconductor materials are n-type semiconductors. If majority carriers are holes, such extrinsic semiconductor materials are p-type semiconductors. Extrinsic semiconductors are the basic materials for luminescent devices.

4.1.1.3 p–n junction

Through the proper process, the conduction types in different regions of semiconductor single crystal materials are n-type and p-type, respectively, and the junction of the two forms the p–n junction. When the p–n junction is formed, due to the difference of carrier concentration in the n zone and the p zone, the majority carrier electrons in the n zone and the majority carrier holes in the p region diffuse to the other region and recombine with their majority carriers, respectively. This results in a decrease in the electron concentration near the n zone side of the p–n junction, leaving an immobile donor ion, resulting in a localized positive charge zone; the hole concentration near the p zone side of the p–n junction decreases, leaving an immobile acceptor ion, resulting in a localized negative charge region. Since the local positive and negative charge region exists, a built-in electric field from the n zone to the p zone will be generated near the p–n junction. The electric field obstructs the continued diffusion of electrons from the n zone to the p zone, while allowing the minority carrier holes in the n region to drift to the p region. Similarly, the electric field obstructs the holes in the p region continuing to diffuse to the n region, while allowing the minority carrier electrons in the region to drift to the n region. With the weakening of diffusion and the enhancement of drift, the dynamic balance of carriers is finally realized.

A region in which carriers are depleted near the p–n junction is called the space-charge region or the depletion region. As a whole, the space-charge region is electrically neutral, and the p–n junction is in an unbalanced state under the condition of applied voltage. The p zone of the p–n junction is connected to the positive electrode of the power supply. The n zone is connected to the negative electrode, and the p–n junction is in the forward bias state. At this time, the electric field generated by the applied voltage in the space-charge region is opposite to the self-built electric field, and the diffusion motion of the carrier is strengthened. Because most carriers are involved in the diffusion motion, a large forward current will be formed. If the p region is connected with the negative electrode of the power supply, the n region is connected with the positive electrode of the power supply, the applied voltage produces the same electric field as the self-built electric field in the space-charge region, the diffusion motion is weakened, and the drift motion of the minority carriers in the depletion region is strengthened. Since it is a minority carrier, the formed reverse current is small, and it can be considered that the p–n junction is in the cutoff state.

For most semiconductor electronic devices and optoelectronic devices, the core part is the p–n junction, so mastering the basic structure and properties of the p–n junction (shown in Fig. 4.2) can help to understand the working principle of these devices.

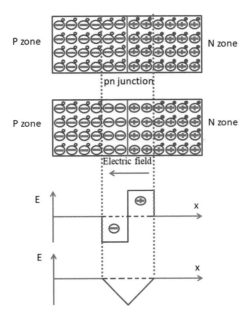

Figure 4.2 The properties of p–n junctions.

4.1.1.4 Direct and indirect bandgap semiconductors

The semiconductor crystals have two band structures: direct bandgap and indirect bandgap. As shown in Figs. 4.3 and 4.4, if the wave vector k position at the bottom of the conduction band is the same as at the top of the valence band, the corresponding bandgap is a direct bandgap. If the wave vector k position of the conduction band bottom is different from that of the valence band top, the corresponding bandgap is indirect bandgap. Correspondingly, semiconductors are also divided into direct bandgap semiconductors and indirect bandgap semiconductors, which show large differences in electrical and optical properties. Direct bandgap semiconductor materials are usually used to make luminescent devices, while indirect bandgap semiconductor materials are mainly used in photodetectors.

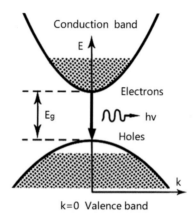

Figure 4.3 Scheme of direct bandgap.

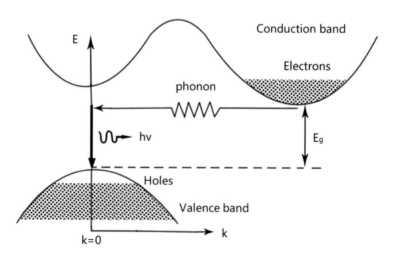

Figure 4.4 Scheme of indirect bandgap.

4.1.2 Semiconductor Luminescence

While electrons in semiconductor materials transition from high energy state to low energy state, they release excess energy in the form of photons, which is called radiative transition. The process of radiative transition is also the luminescence process of semiconductor materials. According to the different excitation modes, the luminescence mechanism of semiconductor materials can be divided into photoluminescence and electroluminescence. Photoluminescence is the re-emission process of light after semiconductor materials absorb higher energy photons. Electroluminescence is a process of light emission caused by current excitation in semiconductor materials. Whether photoluminescence or electroluminescence, the process of radiative transition to produce photons is shown in Fig. 4.5. When electrons transition from a high energy state (conduction band) to a low energy state (valence band), photons with corresponding energy intervals are produced. The photon wavelength is

$$\lambda \approx \frac{1.240}{E_g} \mu m \tag{4.2}$$

The unit of E_g is eV. Assuming that the number of electrons in the high energy state is N, the spontaneous emission recombination process can be described by formula (4.3), that is,

$$\left(\frac{dN}{dt}\right)_{radiative} = -AN \tag{4.3}$$

Clearly, the spontaneous emission rate is determined by Einstein A coefficient. The number of photons emitted at a given time is positively proportional to the number of electrons in a high energy state, that is,

$$N(t) = N(0)\exp(-At) = N(0)\exp(-t/\tau_R) \tag{4.4}$$

The τ_R in the formula is radiation life: $\tau_R = A^{-1}$, which is high energy state radiation life in addition to radiation transition. If the nonradiative transition process is faster than the radiative transition process, only a small amount of light is emitted. When both the radiative and nonradiative transitions are considered, there are

$$\left(\frac{dN}{dt}\right)_{total} = -\frac{N}{\tau_R} - \frac{N}{\tau_{NR}} = -N\left(\frac{1}{\tau_R} + \frac{1}{\tau_{NR}}\right) \tag{4.5}$$

A nonradiative lifetime is τ_{NR} in the formula. Considering the luminous efficiency, there are

$$\eta_R = \frac{N/\tau_R}{N(1/\tau_R + 1/\tau_{NR})} = \frac{1}{1+\tau_R/\tau_{NR}} \qquad (4.6)$$

The emitted light η_R close to 1 if $\tau_R \ll \tau_{NR}$; the emitted light becomes very small and the luminescence efficiency is low if $\tau_R \gg \tau_{NR}$. Therefore, high-efficiency luminescent devices require much less radiation life than nonradiation life.

The luminescence process of semiconductor is complicated. It is related to the delay mechanism of energy in semiconductor. The shape of emission spectrum is affected by the thermal distribution of electrons and holes in the band.

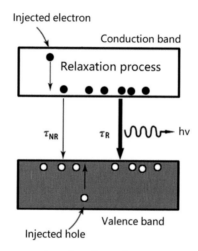

Figure 4.5 Basic process of semiconductor luminescence.

4.1.2.1 Luminescence process of direct bandgap semiconductor materials

The electrons at the bottom of the conduction band transition to the top of the valence band, recombining with the holes. The recombination process satisfies the conservation laws of energy and momentum. Hence, the bandgap energy E is called photon energy. For the recombination process of semiconductors, the momentum of radiant photons is much smaller than that of electrons, so in the recombination process, the photon momentum is negligible. It is considered that the momentum of electrons does not change in the direct transition, that is, the k is invariant, as shown in Fig. 4.3.

4.1.2.2 Luminescence process of indirect bandgap semiconductor materials

For indirect recombination, it is also necessary to satisfy the conservation laws of energy and momentum. Because the conduction band bottom corresponds to the valence band top with different k, phonon participation is required in the recombination process. Assuming that the phonon energy is E, then the photon energy is square $hv = E_g \pm E_p$, "+" indicating the absorption of a phonon, and "–" means the emission of a phonon, as shown in Fig. 4.4.

4.2 Fluorescence Properties of Carbon Quantum Dots

4.2.1 Introduction of Carbon Quantum Dots

CQDs are typically referred to as a class of zero-dimensional (0D) carbon nanoparticles with sizes below 10 nm, whose electronic bandgap structures are largely influenced by the remarkable quantum confinement effect. Since their first discovery in 2006, CQDs have attracted intensive attention due to their unique advantages compared with fluorescent dye molecules and conventional semiconductor quantum dots, such as high luminescence, biocompatibility, water solubility, excellent photostability, and low toxicity. Generally, CQDs prepared from solution are indigenously functionalized with abundant surface groups, especially oxygen-related functional groups, such as carboxyl and hydroxyl, which impart excellent water solubility and suitable chemically reactive groups for surface passivation and functionalization. More importantly, the optical properties such as tunable photoluminescence ranging from deep ultraviolet to near-infrared (NIR) and exceptionally efficient multiphoton up-conversion of CQDs can be tuned by their size, shape, surface functional groups, and heteroatom doping. These intriguing properties afford CQDs a variety of potential biological applications such as cell imaging, cancer therapy, and drug delivery. In 2010, Pan et al. successfully cut grapheme sheets into blue luminescent CQDs via a hydrothermal route, expanding the application of graphene-based materials to optoelectronics such as photovoltaic devices, LEDs, photodetectors, and photocatalysis. In 2013, a facile and high-output method for the fabrication of CQDs with a quantum yield (QY) as high as ca. 80% was developed, marking a significant advance in this area. However, most luminescent CQDs show obvious excitation-dependent fluorescence originating from surface defects serving as capture centers for excitons, which results in fundamental limitations to the effectiveness of carrier injection for optoelectronic applications. Notably, we have first demonstrated monochrome electroluminescent LEDs based on bright multicolor bandgap fluorescent CQDs directly as an active emission layer. Furthermore, we have demonstrated multicolored narrow-bandwidth emission (full width at half maximum of 30 nm) from triangular CQDs for high color-purity full-color LEDs, opening up the great prospects of CQDs for the next-generation display technology. In addition, CQD-based random lasers and wide-color gamut backlight displays have been first realized by our group as well.

Triplet-excited-state-involved materials have attracted tremendous attention for their higher electroluminescence efficiency than their fluorescent counterparts owing to the allowable utilization of the three quarters of the electrically generated excitons for light emission. Surprisingly, novel room temperature phosphorescence (RTP) and thermally activated delayed fluorescence (TADF) properties have been discovered by embedding CQDs into various matrices, including zeolites, polyvinyl alcohol, polyurethane, potassium aluminum sulfate, and recrystallized urea/biuret. Moreover, CQDs with intrinsic RTP properties without the need for additional matrix compositing have also been reported recently, demonstrating promising alternative RTP materials based on the CQDs for highly efficient optoelectronic devices.

There have been already a number of excellent papers focusing on different aspects of CQDs, such as their synthesis, surface functionalization, photoluminescence properties, and biological applications. In the following sections, we will primarily update the latest significant developments in intriguing luminescence properties and mechanisms such as RTP and TADF, and then the key challenges, feasible improvements, and perspectives of CQD-based optoelectronics for future developments are highlighted.

4.2.2 Fluorescence Emission from Bandgap Transitions of Conjugated π-Domains

CQDs are a new class of nanomaterials that have attracted significant attention in the past decade. Two popular models have been proposed to explain the fluorescence mechanism of CQDs even though the exact origin of their fluorescence emission remains debatable, and more research is needed in order to paint a clearer picture of the mechanism of their fluorescence emission. One is based on the bandgap emission in the conjugated π-domains, while the other is related to surface defect states, which primarily manifest the edge effect.

The quantum confinement effect (QCE) is the major feature of CQDs, which occurs when CQDs are smaller than their exciton Bohr radius. For CQDs with an intact carbon core and fewer surface chemical groups, the bandgap transition of conjugated π-domains is thought to be the major intrinsic PL center. Density functional theory (DFT) calculations have clearly demonstrated that the bandgaps of CQDs decrease approximately from 7 eV for benzene to 2 eV for CQDs consisting of 20 aromatic rings (Fig. 4.6a). Using DFT and time-dependent DFT calculations, Sk et al. also found out that the PL of CQDs originates from the QCE of π-conjugated electrons in the sp^2 atomic framework and can be tuned by their size, edge configuration, and shape. Typically, the localized states at zig-zag-edged CQDs lower the energy of the conduction band and thus reduce the bandgap, which is distinctive compared with the arm-chair-edged CQDs. Therefore, it is expected that the similarly sized CQDs with a zig-zag edge would narrow the bandgap and consequently red shift the fluorescence emission.

Fan's group successfully synthesized red fluorescent CQDs with no chemical modification by direct electrochemical exfoliation of graphene in a $K_2S_2O_8$ solution and identified isolated sp^2 domains with a diameter of 3 nm (Fig. 4.6b,c). This is the first report on the direct observation of "molecular" sp^2 domains with a diameter of about 3 nm that are directly responsible for the red emission. Furthermore, the size-dependent QCE of CQDs has been observed by our group. They reported the first demonstration of bright multicolor bandgap fluorescent CQDs (MCBF-CQDs) from blue to red with a QY up to 75% for blue fluorescence (Fig. 4.7a). The MCBF-CQDs show strong excitonic absorption bands in the UV-vis absorption spectra (Fig. 4.7b) centered at about 350, 390, 415, 480, and 500 nm for B-, G-, Y-, O-, and R-BF-CQDs, respectively, which is similar to the absorption characteristics of quantum-confined semiconductor QDs but quite different from that of previously reported CDs within only UV region. The fluorescence peaks of MCBF-CQDs are centered at about 430 (B-), 513 (G-), 535 (Y-), 565 (O-), and 604 nm (R-BF-CQDs) (Fig. 4.7c). The red-shifted fluorescence colors from blue to red are well consistent with the red-shifted excitonic absorption bands, manifesting the band edge nature of the optical transitions in the MCBF-CQDs. Another piece of evidence for the bandgap emission properties was obtained by time-resolved PL analysis (Fig. 4.7d), which revealed a monoexponential decay of about 14.2, 12.2, 11.3, 8.8, and 6.8 ns for B-, G-, Y-, O-, and R-BF-CQDs, respectively. The monoexponential decay characteristics indicate the band edge exciton-state decay rather than the defect-state decay within the MCBF-CQDs, which is conducive to efficient fluorescence emission and visibly different from that reported for CDs with multi-exponential decay. Meanwhile, the decreased highest occupied molecular orbital (HOMO) levels from 5.72 to 5.27 eV determined by means of ultraviolet photoelectron spectroscopy (UPS) and the increased lowest unoccupied molecular orbital (LUMO) levels from 2.70 to 3.15 eV (Fig. 4.7e) directly reveal the bandgap transitions in MCBF-CQDs. More recently, researchers have reported the synthesis of multicolored, high color-purity, and narrow-bandwidth (full width at half maximum of 29–30 nm) emission triangular CQDs (NBE-T-CQDs) with a QY up to 54–72% (Fig. 4.8). For the first time, we overturn the belief that CQDs can only give broad emission and inferior color purity with the full width at half maximum commonly exceeding 80 nm, and achieve an unprecedented narrow bandwidth of 29 nm. The typical aberration-corrected high-angle annular dark-field scanning transmission electron microscopy (HAADF-STEM) images of the NBE-T-CQDs demonstrate the almost defect-free graphene crystalline structure with an obvious triangular shape.

Other detailed structural characterization methods demonstrated that the as-prepared NBE-T-CQDs are highly crystalline and have a unique triangular structure functionalized with pure electron-donating hydroxyl groups at the edge sites. More importantly, the femtosecond transient absorption spectroscopy and temperature-dependent emission narrowing were conducted to reveal the emission mechanism of NBE-T-CQDs, which demonstrated the dramatically reduced electron–phonon coupling due to triangular structural rigidity as the origin of the high color-purity excitonic emission. In addition, the DFT calculations revealed that the unique triangular CQDs showed highly delocalized charges and high structural stability, which subsequently resulted in dramatically reduced electron–phonon coupling and further led to the high color-purity bandgap emission (Fig. 4.9).

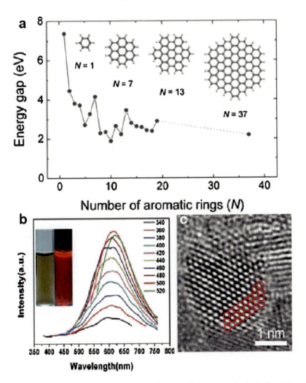

Figure 4.6 (a) Energy gap of $\pi - \pi^*$ transitions calculated using DFT as a function of the number of fused aromatic rings. (b) PL spectra and (c) sp^2 domains of RF-CQDs. (Yuan et al., 2016; Yuan et al., 2019).

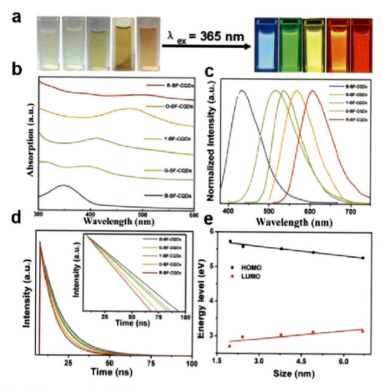

Figure 4.7 (a) Photographs of MCBF-CQDs under daylight (left) and fluorescence images (right) under UV light (excited at 365 nm). (b) UV-vis absorption, (c) normalized fluorescence, and (d) time-resolved PL spectra of B-BF-CQDs, G-BF-CQDs, Y-BF-CQDs, O-BF-CQDs, and R-BF-CQDs. Dependence of (e) the HOMO and LUMO energy levels with respect to the size of MCBF-CQDs.

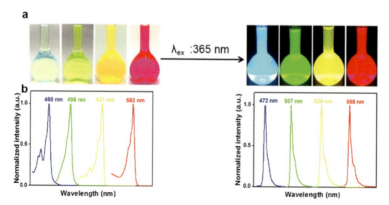

Figure 4.8 (a) Photographs of the NBE-T-CQDs in ethanol solution under daylight (left) and fluorescence images under UV light (excited at 365 nm) (right). (b) The normalized UV-vis absorption (left) and PL (right) spectra of B-, G-, Y-, and R-NBE-T-CQDs, respectively.

Fluorescence Properties of Carbon Quantum Dots | **105**

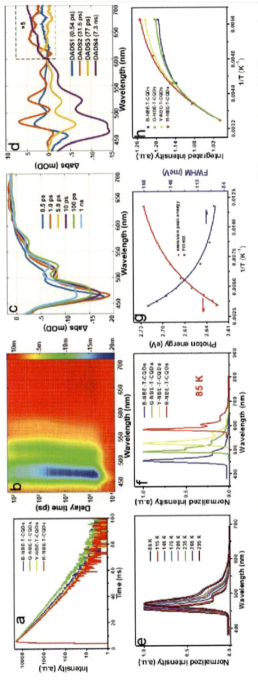

Figure 4.9 Ultrafast dynamics of the photoexcited states and temperature-dependent PL spectra of the NBE-T-CQDs. (a) Time-resolved PL spectra of the NBE-T-CQDs. (b) Two-dimensional pseudo-color map of TA spectra of B-NBE-T-CQDs expressed in ΔOD (the change in the absorption intensity of the sample after excitation) as a function of both delay time and probe wavelength excited at 400 nm. (c) TA spectra of B-NBE-T-CQDs at indicated delay times from 0.5 ps to 1 ns. (d) Results of the global fitting with four exponent decay functions. (e) The normalized temperature-dependent PL spectra of B-NBE-T-CQDs. (f) The normalized PL spectra of the NBE-T-CQDs acquired at 85 K. (g) The plots of the emission peak energy and full width half maximum of B-NBE-T-CQDs as a function of temperature (85 to 295 K). (h) The plots of the integrated PL emission intensity of the NBE-T-CQDs as a function of temperature (175 to 295 K).

4.2.3 Fluorescence Emission of Surface Defect-Derived Origin

The second kind of the fluorescence mechanism of CQDs arises from surface-related defect states. Both sp^3- and sp^2-hybridized carbons and other surface defects of CQDs, such as oxygen-containing functional groups, can serve as capture centers for excitons, thus giving rise to surface-related defect-state fluorescence. With support from DFT calculations, it has been demonstrated that the carboxyl groups on the sp^2-hybridized carbons could induce significant local distortions and consequently narrow the energy gap. And the optical properties may be completely changed such as radically different fluorescence emission bands and intensity distributions after the reduction in these oxygen-containing functional groups (Fig. 4.10). Hu et al. prepared a series of CQDs by varying the reagents and reaction conditions and concluded that the surface epoxides or hydroxyls were predominantly responsible for the resulting PL red shift. The fluorescence intensity and even the peak position of CQDs originating from surface-related defect states can be changed significantly with the pH of their solution, which may be ascribed to the protonation or deprotonation of the oxygen-containing functional groups and thus shift the Fermi level of the CQDs. Our group has reported novel multicolor fluorescent CQDs, which are responsive to all pH from 1 to 14 and can be observed even with the naked eye. A novel quinone structure in the CQDs, which was for the first time transformed from the lactone structure under strong alkaline conditions, is responsible for the red emission under strong alkaline conditions. The pH-dependent PL is very important for exploring the fluorescence mechanism of CQDs, the understanding of which is still at an early stage and much deeper research is still required to resolve this issue. In addition to the oxygen-containing functional groups, amine-containing groups also contribute to the surface defect states in CQDs, which is attributed to their unpaired electrons as electron donors, and improve the fluorescence properties of CQDs. Furthermore, the surface-defect-state emission can be due to surface modification of CQDs. For instance, after the passivation of –COOH and epoxy of CQDs into –CONHR and –CNHR with alkylamines, the intrinsic blue emission was greatly enhanced, while the surface-defect-state green emission disappeared. It should be emphasized that the obvious excitation-dependent fluorescence of the above reported CQDs dominated by surface defects severely results in fundamental limitations to the effectiveness of carrier injection for optoelectronic applications.

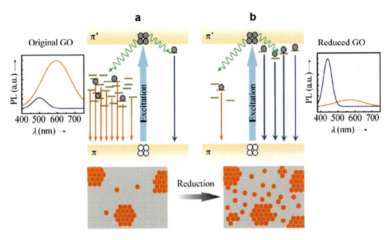

Figure 4.10 Proposed PL emission mechanisms of (a) the predominant "red emission" in GO from disorder-induced localized states and (b) the predominant "blue emission" in reduced GO from confined cluster states.

4.2.4 Up-Conversion Fluorescence

In addition to the conventional down-conversion fluorescence, certain CQDs show up-conversion fluorescence emission (anti-Stokes emission), which is orders of magnitude higher in efficiency compared to conventional fluorophores. Up-conversion fluorescence is particularly attractive for in vivo imaging due to the deep-tissue penetration ability of long excitation wavelength especially in the NIR region. Our group has developed the first facile and large-scale synthesis of nitrogen-rich CQDs (NRCQDs) with bright two-photon fluorescence (TPF). And we further demonstrated that a large imaging depth of up to 440 mm for live rat liver tissues could be achieved by NRCQD-based TPF imaging, which indicates that CQDs with TPF possess a further application future.

Although great progress regarding the fluorescence properties of CQDs has been achieved, there are still some issues which are imperative to be solved, such as the still unclear fluorescence mechanisms and their lower QYs in the long-wavelength region, especially in the NIR region. Furthermore, both in-depth experimental verifications and theoretical calculations are still urgently needed to provide convincing explanations for the fluorescence mechanism. One of the most important challenges to demonstrate the fluorescence mechanism of CQDs is to develop an efficient and controllable method for producing high-quality CQDs with a well-defined structure. Most CQDs reported to date usually show different sizes and different crystallinity, numerous but uncertain surface defects as well as functional groups, making it difficult to distinguish the key factors that influence the fluorescence emission. Once the structures of CQDs are well controlled by suitable methods, it will be possible to make the fluorescence mechanisms clear.

4.3 Room Temperature Phosphorescence Properties of CQDs

The quantum mechanically allowed transition of singlet excitons (S_1) with antisymmetric spins and a total spin quantum number of zero (S = 0) to the ground state (S_0) results in fluorescence within nanoseconds. Conversely, the quantum mechanically forbidden transition of the triplet state (T_1) with even symmetry and S = 1 to the singlet ground state results in phosphorescence with lifetimes in the microseconds to seconds regime. Because of the corresponding multiplicities of the angular momentum states (i.e., m_S = 0 for S = 0 and m_S = −1, 0, 1 for S = 1) and the random nature of spin production in electroluminescent devices, exciton formation under electrical excitation typically results in 25% singlet excitons and 75% triplet excitons. In general, the theoretical maximum external quantum efficiency (ZEQE) for electroluminescent LEDs as a key parameter can be estimated by Eq. (4.7).

$$\eta_{EQE} = \eta_{int} \times \eta_{out} = (\gamma \times \eta_\gamma \times \Phi_{PL}) \times \eta_{out} \qquad (4.7)$$

where η_{int} is the internal quantum efficiency and η_{out} is the light-outcoupling efficiency (typically $\eta_{out} \approx 0.2$); γ is the charge balance of injected holes and electrons (ideally γ = 1); η_γ is the efficiency of radiative exciton production (η_γ = 0.25 for conventional fluorescent emitters); and Φ_{PL} is the photoluminescence (PL) QY of the emitter material. For fluorescence materials, 75% of the electrically generated energy is dissipated as heat by triplet excitons owing to the spin-forbidden nature of triplet-state transitions, leading to the theoretically highest external quantum efficiency of 5% after considering a light-outcoupling efficiency of 20% in the device. In contrast, the phosphorescent materials, due to their ability to harvest light from both triplet and singlet excitons, allow the internal quantum efficiency of the device to reach nearly 100% (Fig. 4.11) and have thus aroused considerable interest in these materials for optoelectronic applications. Generally, short lifetimes (dozens of microseconds), in addition to high phosphorescence quantum efficiencies, are required for phosphorescent materials to realize high-performance electroluminescent devices. This is because the long lifetime of triplet states will inevitably result in a severe efficiency roll-off at high current densities in electroluminescence due to triplet–triplet exciton annihilation (TTA) or triplet–polaron annihilation (TPA).

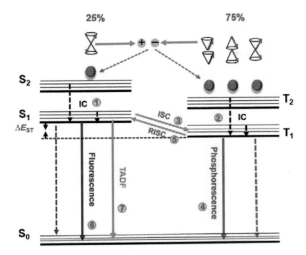

Figure 4.11 Electroluminescence processes in phosphorescent and TADF materials. The transition process of electrically generated excitons for fluorescence (A–B), phosphorescence (A–B–C–D), and TADF (B–E–Q). A and B: internal conversion (IC), C: intersystem crossing (ISC), D: phosphorescence, E: reversible intersystem crossing (RISC), F: fluorescence.

So far, luminogens with phosphorescence features have been typically limited to inorganics or organometallic complexes, and they often suffered from poor processability, high cost, and heavy metal toxicity. In contrast, recently, metal-free RTP materials, especially pure organic RTP phosphors, have attracted much attention as advanced phosphorescent luminogens on account of their attractive characteristics such as good processability, low cost, versatile molecular design, and facile functionalization. However, metal-free RTP materials are rare and generally exhibit dim phosphorescence under ambient conditions because of inefficient spin–orbit coupling, the long-lived sensitive triplet excitons and quenching by impurity with long lifetimes. To achieve metal-free RTP luminogens, first we need to introduce heavy atoms and heteroatoms (N, O, S, P, and so on) into the luminescent skeletons to facilitate the effective intersystem crossing and meanwhile to modulate the aggregation behaviors to suppress the nonradiative dissipation by polymer aggregation, hydrogen bonds, or supramolecular assembly. Huang et al. proposed that H-aggregation stabilized the triplet excitons by enhancing the intersystem crossing process, in pursuit of ultralong RTP. Tang et al. suggested that the crystallization-induced phosphorescence mechanism might be mainly responsible for the RTP character because of its effective inhibition of nonradiative decay. Chi et al. identified the key role of the intermolecular electronic coupling in achieving efficient persistent RTP. However, most organic RTP materials often require ordered crystal structures or stringent conditions associated with inert atmospheres to suppress the nonradiative relaxation process of the triplet states, which inevitably results in quenching of phosphorescence in the amorphous form and fundamentally limits its practical application.

As an emerging class of luminescent nanomaterials, novel RTP properties of CQDs have been expected, which offers an intriguing prospect for utilizing triplet states for breaking through the 5% limitation of the traditional fluorescence devices. However, the search for CQD-based RTP materials remains a challenging research area because of the weak spin–orbit coupling of singlet to triplet. In 2013, Zhao et al. first reported pure organic RTP by dispersing CQDs in the polyvinyl alcohol (PVA) matrix. This RTP emission is proposed to originate from triplet excitons of aromatic carbonyls on the surface of CQDs and the PVA molecules, which effectively protect their energy from rotational or vibrational loss by rigidifying the groups with hydrogen bonds. Since then, further studies have shown that the layered double hydroxides, cyanuric acid, and inorganic crystalline nanocomposites are also applicable as host materials for CQDs to exhibit remarkable RTP upon excitation by UV light, which indicates the potential of intersystem crossing and lays a solid foundation for the progress of promising alternative metal-free RTP materials based on the CQDs. However, this strategy using CQD composites to achieve RTP is more vulnerable to phase segregation, poor thermal stability and conductivity, which severely impedes their practical applications. Therefore, single-component CQDs with intrinsic RTP properties are highly desired. Blue–green RTP CQDs without the need for additional matrix compositing were developed through hydrothermal treatment of polyacrylic acid and ethylenediamine (Fig. 4.12a–c). This study indicated that the covalent crosslinking occurring in the interior of CQDs can generate luminescence centers while simultaneously restricting their vibration and rotation, thus providing favorable conditions for effective intersystem crossing. Before long, other examples including a gram-scale method for the preparation of ultralong RTP CQDs (1.46 s) via microwave-assisted heating of ethanolamine and phosphoric acid aqueous, and blue–green self-protective RTP emission based on fluorine and nitrogen codoped carbon dots (FNCDs) synthesized from glucose and $Et_3N·3HF$, further demonstrate the intrinsic ability of intersystem crossing in CQDs. However, the reported RTP emission wavelengths of those CQDs reported earlier were constrained to the blue or green region below 540 nm, and color-tunable RTP emission is of great necessity and significance for optoelectronic applications. CQD-based color-tunable fluorescence and RTP over the broad spectral range of 500–600 nm were revealed by the seeded growth method (Fig. 4.12d–f). The origin of RTP is believed to be related to the surface nitrogen-containing groups of the CQDs, with a contribution from hydrogen bonds to protect the triplet states from quenching, and the additional stabilizing action of the PVP polymer chains.

112 | *Carbon Quantum Dot Luminescent Materials*

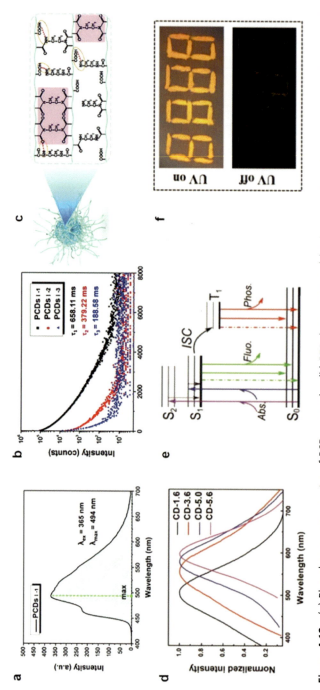

Figure 4.12 (a) Phosphorescence spectra of PCDs$_{I-1}$ powder. (b) RTP decay spectra. (c) Schematic illustration of the crosslink-enhanced emission effect. (d) Normalized RTP emission spectra of the four CQD powders mentioned on the frame. (e) Jablonski diagram for a three-level system illustrating possible PL and RTP pathways for CQDs. (f) Images of luminescent characters written using CQD-5,6 and an orange fluorescent pen on the non-luminescent background paper under the conditions of UV light on and off.

4.4 Thermally Activated Delayed Fluorescence Properties of CQDs

When T_1 and S_1 are close in energy, i.e., the singlet–triplet energy splitting (ΔE_{ST}) is small, the endothermic reverse intersystem crossing (RISC) process can be overcome at high temperature. As a result, the nonradiative triplet excitons, due to a spin forbidden T_1–S_0 transition, are transformed to singlet excitons via RISC, leading to TADF emission (S_1–S_0) with lifetimes in the range of hundreds of nanoseconds to dozens of milliseconds (Fig. 4.11). Thus, the TADF materials also can employ the triplet-state energy for improving the energy conversion efficiency. Recently, Yu et al. reported a facile and general "dots-in-zeolites" strategy to in situ confine CQDs in zeolitic matrices during hydrothermal/solvothermal crystallization to generate high-efficient TADF. The resultant CQDs@zeolite composites exhibit high QY up to 52.14% and ultralong lifetimes up to 350 ms at ambient temperature and atmosphere (Fig. 4.13), which introduces a new perspective to develop CQD-based TADF materials for advanced optoelectronic device applications.

Despite the intensive work on the RTP and TADF properties of CQDs, the research on CQD-based triplet-state modulation is still at an early stage with extremely low phosphorescence quantum efficiencies and long lifetime (hundreds of milliseconds) compared with well-developed transition-metal complexes. Besides, the color range of CQD-based RTP is quite narrow, leaving long-wavelength emission (particularly orange or red emission) unexplored. In addition, the luminescence mechanism of the CQD-based RTP is not fully understood until now due to the complicated and unclear structural features of CQDs. Therefore, to obtain efficient CQD-based RTP, first we need to populate triplet excitons by introducing heteroatoms to enhance intersystem crossing and meanwhile suppress the nonradiative dissipation for achieving its application in electroluminescence. Furthermore, more efforts should be devoted to advanced and high-resolution characterization techniques as well as to computational simulations, which may increase our understanding of structure–property relationships and the fundamental phosphorescence mechanisms. Related works along this line are underway in our laboratory and will be reported in due course.

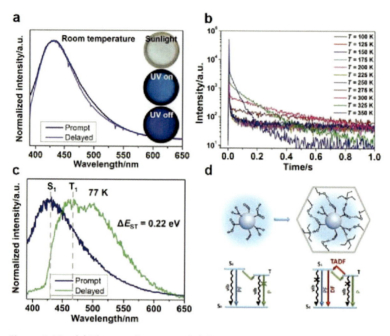

Figure 4.13 (a) The steady-state and delayed photoluminescence spectra of CDs@AlPO-5 excited under 370 nm at room temperature (inset: photographs under sunlight, UV lamp of 365 nm, and UV turned off). (b) Temperature-dependent transient photoluminescence decay. (c) The steady-state photoluminescence spectrum (deep blue line) and delayed photoluminescence spectrum (olive line) excited under 370 nm at 77 K. (d) TADF mechanism of CDs@zeolite composites.

4.5 Summary

In summary, we have reviewed the recent advances in optical properties and luminescence mechanisms in CQDs. There is no doubt that the development of CQDs for optoelectronic applications has seen great progress within a short time. However, the research still faces a set of challenges in terms of fundamental understanding and practical applications as stipulated below.

First, one of the most controversial fundamental issues of CQDs is the fluorescence mechanism. Further new structural characterization methods and theoretical calculations are badly desired to provide a clearer coherent picture. Moreover, CQDs with a high QY comparable to traditional semiconductor QDs at longer wavelengths beyond the blue light region are yet to be synthesized on a large scale through facile methods.

Second, triplet-excited-state-involved materials have attracted considerable attention as a promising and efficient approach to harvest triplet excitons for electroluminescence applications. However, CQD-based triplet-state emission, especially the RTP properties of CQDs, is rare and generally exhibits extremely dim RTP with long lifetime due to the weak spin–orbit coupling of singlet to triplet. Furthermore, the CQD-based RTP with long-wavelength emission (orange or red) remains underexplored. Thus, the practical applications of CQDs with RTP properties are only limited to anti-counterfeiting, optical recording, sensors, and security protection, and there remains a long way to go through potential applications in electroluminescence. Therefore, new strategies for production of CQD-based triplet-state emission with high quantum efficiencies and short lifetime need to be urgently developed.

Third, with continual achievements of the multicolor PL and high QY, CQDs will become more and more promising for optoelectronic applications. For phosphor-based white LEDs (WLEDs), developing high-performance single-component warm WLEDs based on intrinsically broadband white emissive CQDs covering the entire visible spectral window from 400 to 700 nm is highly desired for solid-state lighting, which has so far remained a huge challenge.

Fourth, for electroluminescence, the successful synthesis of unprecedented narrow-bandwidth emission triangular CQDs will set the stage for developing next-generation CQD-based display technology. However, some problems still should be settled for further development of full-color displays. For example, new strategies for preparing narrow-bandwidth CQDs ranging from blue to NIR with high QY need to be developed. Furthermore, efficient RTP or TADF CQDs as well as optimization of the charge injection with better controlled device fabrication are highly demanded for improving the performance of CQD-based electroluminescent LEDs. In addition, facile methods for preparing CQDs with low costs and on a large scale need to be developed for practical applications.

Finally, the research on CQDs in the field of laser, solar cells, and photodetectors is just the beginning, and there is a lot of room for further development. With the continued advances of experimental and theoretical studies, we can expect to achieve more groundbreaking advances in both fundamental research and optoelectronic applications of CQDs in the future.

Questions

1. What are the characteristics of the space-charge region of a p–n junction? How is it formed?
2. What are the differences between the luminescence processes of direct and indirect bandgap semiconductors?
3. Can you list some light-emission mechanisms of carbon quantum dots?
4. Explain the up-conversion fluorescence mechanism of carbon quantum dots.
5. What is the mechanism of room temperature phosphorescence of carbon quantum dots?
6. What is the mechanism of delayed fluorescence by thermal activation of carbon quantum dots?

References

Hou H. *Optoelectronic Materials and Devices* (2nd edition), Beihang University Press (2018).

Yuan F., Li S., Fan Z., Meng X., Fan L., and Yang S. "Shining carbon dots: synthesis and biomedical and optoelectronic applications." *Nano Today*, **11**(5), 556–586 (2016).

Yuan T., Meng T., He P., Shi Y., Li Y., Li X., Fan L., and Yang S. "Carbon quantum dots: an emerging material for optoelectronic applications." *J. Mater. Chem. C*, **7**(23), 6820–6835 (2019).

Chapter 5

Application in Lithium-Ion Battery

Lithium-ion batteries (LIBs) are widely used in portable appliances such as laptop computers, cameras, and mobile communications because of their unique performance advantages. Large-capacity LIBs are expected to be one of the main power sources for electric vehicles in our sustainable society, as well as in artificial satellites, aerospace, and energy storage. With the shortage of energy and the pressure of environmental protection in the world, LIBs are becoming more and more popular to consumers. Especially with the emergence of new materials, the development and application of LIBs in industry have been accelerated. In this chapter, new materials with hierarchical porous micro/nanostructures are taken as a window to gain an idea about the material and application aspects of LIB research.

Hierarchically porous materials generally present multiple-level porous structures in which the pore length scales range from micropores (<2 nm) to mesopores (2–50 nm) and macropores (>50 nm). More specifically, these materials always consist of assembled molecular units or their aggregates that are embedded in or intertwined with other units or aggregates, which may, in turn, be similarly organized at increasing size levels. These aesthetic architectures enable the structured materials to obtain unique properties and functionalities.

Materials and Interfaces for Clean Energy
Shihe Yang and Yongfu Qiu
Copyright © 2022 Jenny Stanford Publishing Pte. Ltd.
ISBN 978-981-4877-66-4 (Hardcover), 978-1-003-14223-2 (eBook)
www.jennystanford.com

Hierarchically porous micro-/nanostructured materials can enhance the electrochemical performance (e.g., rate capability and cycling stability) of LIBs. The advantages of these materials can be explained as follows. First, the transport of both electrons and lithium ions is greatly affected by the sizes of the nanomaterials. The ion diffusion in solid-state electrode materials is directly related to the transport path, as expressed by Eq. (5.1):

$$\tau = L^2/D \qquad (5.1)$$

where L, D, and τ represent the diffusion length, diffusion coefficient, and diffusion time of Li$^+$, respectively. Therefore, the diffusion length of the lithium ion in nanocrystalline grains within hierarchically porous structured electrode materials can be reduced, resulting in improved rate capabilities. Second, hierarchically porous micro-/nanostructured electrode materials possess relatively high specific surface areas, which could increase the contact area between the electrode and the electrolyte, and thereby enhance utilization of the active materials and increase the gravimetric capacity. Third, the unique structure could offer enough space to accommodate the volume change during the charge/discharge process, giving rise to better cyclic stability.

In the following sections, we first summarize the synthesis of hierarchically porous micro-/nanostructures via morphology-conserved transformation, and then discuss their applications in LIBs.

5.1 Various Precursors to Hierarchically Porous Micro-/Nanostructures

Hierarchically porous structured materials have many exceptional properties, such as ultrahigh surface areas, controlled pore sizes and shapes, and nanoscale effects, making them the most promising candidates for energy conversion and storage device applications. However, how to control the structures, sizes, and shapes of the materials remains challenging in current material syntheses. Fortunately, the conservation of morphology from metal-based precursors to porous metal oxides is an effective method compared with intricate template approaches. Electrospinning is an example of morphology-conserving transformation methods for producing hierarchically porous micro-/nanofibers in different forms, such as core–shell hollow porous micro-/nanofibers. Given the broad scope of the topic, the following sections will focus on select works utilizing some interesting precursors.

5.1.1 Metal Hydroxide Precursors

The most common metal-based precursors are metal hydroxides, such as $Mn(OH)_2$, $Co(OH)_2$, $Ni(OH)_2$, $Sn(OH)_2$, $Ce(OH)_2$, $La(OH)_3$, $In(OH)_3$, $Y(OH)_3$, and Ni–Co bimetallic hydroxides. In the synthesis of metal hydroxides, sodium hydroxide, ammonium hydroxide, and hydrazine hydrate are typically used as the hydroxide source. Among them, sodium hydroxide and ammonium hydroxide are the most commonly used sources of insoluble metal hydroxides because their reaction principles are simple. For example, Wang et al. prepared nanocrystalline-assembled bundle-like CuO particles through the calcination of an as-prepared bundle-like $Cu(OH)_2$ precursor. In the typical formation of bundle-like $Cu(OH)_2$, NaOH was directly added to a mixed aqueous solution of $CuCl_2$ and $C_6H_8O_7$. In addition, core–ring structured $NiCo_2O_4$ nanoplatelets were obtained by annealing a mixed precursor of $β-Co(OH)_2$ and $Ni(OH)_2$ at 200°C. The coprecipitation of $Co(OH)_2$ and $Ni(OH)_2$ was also achieved using sodium hydroxide as the precipitant. Needle-like Co_3O_4 nanotubes were successfully converted from a $β-Co(OH)_2$-nanorod precursor, which was produced by reacting $Co(NO_3)_2$ with ammonia solution.

In addition to NaOH and $NH_3·H_2O$, alkaline conditions can also be generated via the thermal decomposition of urea in water, or the hydrolysis of weak acidic salts. Urea has been shown to play a very critical role in the hydrothermal synthesis method. Li et al. prepared a mesoporous ultrathin NiO nanowire network through the thermal decomposition of an $α-Ni(OH)_2$ precursor at 300°C for 2 h in air. The $α-Ni(OH)_2$ precursor was obtained by hydrothermal treatment of $NiCl_2$ in the presence of urea in aqueous solution. Similarly, urchin- and flower-like hierarchical NiO microspheres were successfully synthesized by Pan et al. via calcining $α-Ni(OH)_2$ precursors. The morphologies of the precursors were controlled by simply tuning various sources of nickel salts in the nickel-salt-urea-H_2O ternary system. Anions were observed to intercalate in the crystal lattice of α-Ni(OH)$_2$ and strongly influence the self-assembly process, thereby determining the resulting morphology and structure.

The mechanism by which urea tailored the resulting morphology was proposed as follows: At the appropriate temperature (normally above 80°C), urea gradually hydrolyzes and releases NH$_3$ and CO$_2$, creating weak alkaline conditions for the generation of metal hydroxide or metal carbonate hydroxide nuclei. The growth of nuclei can then be controlled by the release rates of NH$_3$ and CO$_2$, affording sufficient time for the spontaneous self-assembly of nanoblocks to minimize the mutual interaction energy. The related reaction formulas could be written as follows:

For the thermal decomposition of urea in water:

$$CO(NH_2)_2 + 2H_2O \rightarrow 2NH_3 + CO_2 \quad (5.2)$$

$$M_1^{x+} + xNH_3 + xH_2O \rightarrow M_1(OH)_x + xNH_4^+ \quad (5.3)$$

where M_1 refers to metal elements and $M_1(OH)_x$ is insoluble. For the growth of hierarchically structured precursors, except for the decomposition of urea, the application of appropriate weak acid salts (e.g., sodium acetate) is another feasible method to obtain alkaline conditions for precipitating metal ions.

For the hydrolysis of a weak acidic salt using acetate (Ac) salt as an example:

$$M_2(Ac)_y + yH_2O \rightarrow yHAc + M_2^{y+} + yOH^- \quad (5.4)$$

$$M_2^{y+} + yOH^- \rightarrow M_2(OH)_y \quad (5.5)$$

where M_2 refers to metal elements and $M_2(OH)_y$ is insoluble. Hydrolysis of the acetate group resulted in the formation of OH$^-$ (Eq. (5.4)); then the OH$^-$ combined with metal ions to produce metal hydroxides. For instance, nanoporous NiO structures with an actinia shape were fabricated by calcining an α-Ni(OH)$_2$ precursor. In a typical experiment, the α-Ni(OH)$_2$ precursor was obtained hydrothermally by reacting NiCl$_2$ and NaAc in a mixed solution of ethylene glycol (EG) and water. NaAc offered a weakly basic environment for the formation of sheet-like α-Ni(OH)$_2$. In addition, hierarchical NiCo$_2$O$_4$ tetragonal microtubes were fabricated via the thermal transformation of a layered nickel–cobalt-hydroxide precursor, prepared from Ni(Ac)$_2$ and Co(Ac)$_2$ in a solution containing 1,3-propanediol and isopropyl alcohol. Surfactants such as poly(ethylene glycol) (PEG), poly(vinylpyrrolidone) (PVP), and sodium dodecyl sulfate (SDS) have been demonstrated to play important roles in the formation of more complicated micro-/nanomaterials because of their amphiphilic features. For instance, 3D micro-flowery In(OH)$_3$ structures assembled from 2D nanoflakes were prepared via a hydrothermal approach with the assistance of SDS. Additionally, In$_2$O$_3$ retaining its morphology was produced by annealing the In(OH)$_3$ precursor.

5.1.2 Metal Carbonate Precursors

Metal carbonates, such as manganese carbonate, cobalt carbonate, and Co–Mn bimetal carbonate, are another type of precursor that can be used to prepare hierarchically porous structures. When generating metal carbonates, inorganic salts or the decomposition of certain organics is typically used as the source of the carbonate ion. NH$_4$HCO$_3$ and Na$_2$CO$_3$ are the most commonly used inorganic salts to provide carbonate ions for the synthesis of various metal carbonates. For example, LiNi$_{0.5}$Mn$_{1.5}$O$_4$ hollow microspheres were obtained through morphology-conserved synthesis following a three-step route. First, spherical MnCO$_3$ was produced by a simple precipitation method with MnSO$_4$ as the manganese source and NH$_4$HCO$_3$ as the precipitant. Then, the as-formed MnCO$_3$ microspheres were thermally decomposed into porous MnO$_2$, which inherited the microsphere morphology of MnCO$_3$ after the annealing process. Finally, LiOH and Ni(NO$_3$)$_2$ were impregnated into the as-synthesized mesoporous MnO$_2$ microspheres to form LiNi$_{0.5}$Mn$_{1.5}$O$_4$ hollow microspheres. This formation process was analogous to the Kirkendall effect, in which the outward diffusion of Mn and Ni atoms was quicker than the inward diffusion of O atoms. A similar three-step method was applied to prepare LiMn$_2$O$_4$ microspheres. Additionally, a hierarchically porous LiNi$_{1/3}$Co$_{1/3}$Mn$_{1/3}$O$_2$ structure was synthesized by a two-step process: The carbonate precursor (Ni$_{1/3}$Co$_{1/3}$Mn$_{1/3}$)CO$_3$ was prepared using Na$_2$CO$_3$ and NH$_3$HCO$_3$ as precipitants and then mixed with a lithium source to produce the final product.

Various organics can be used to synthesize metal carbonate precursors, but they share certain common features and contain

–CH$_2$OH, –CHO, or –COOH groups. Under a given set of reaction conditions (e.g., hydrothermal), these compounds decompose and then generate CO$_3^{2-}$ anions, which combine with metal cations to form metal carbonate nuclei and subsequently grow to produce the metal carbonate precursor. Typically, fructose and β-cyclodextrin have been used as multifunctional polyol reagents to prepare different MnCO$_3$ microstructures. Hierarchically porous Mn$_2$O$_3$ has been generated via calcining the as-synthesized MnCO$_3$. The proposed mechanism is described in Eqs. (5.6–5.10). The redox reactions between permanganates and poly-based organic molecules (e.g., fructose and β-CD, where CD = cyclodextrin (C$_6$H$_{10}$O$_5$)$_7$) would generate MnCO$_3$ and MnOOH precipitates, which would nucleate during the pyrolysis of polyol-based organic molecules.

$$4\text{R-CH}_2\text{OH} + \text{MnO}_4^- \rightarrow \text{MnOOH} + 4\text{R-CHO} + \text{OH}^- + \text{H}_2\text{O} \quad (5.6)$$

$$\text{R-CHO} + \text{MnO}_4^- \rightarrow \text{MnOOH} + \text{R-COO}^- + \text{H}_2\text{O} \quad (5.7)$$

$$5\text{R-COO}^- + 2\text{MnO}_4^- + 6\text{H}^+ \rightarrow 2\text{Mn}^{2+} + 5\text{CO}_2 + 8\text{H}_2\text{O} \quad (5.8)$$

$$\text{CO}_2 + \text{H}_2\text{O} \rightarrow 2\text{H}^+ + \text{CO}_3^{2-} \quad (5.9)$$

$$\text{Mn}^{2+} + \text{CO}_3^{2-} \rightarrow \text{MnCO}_3 \quad (5.10)$$

Strong oxidants are required for the redox reaction. KMnO$_4$ is among the most commonly used oxidants for the synthesis of manganese-based precursors. Similar syntheses have been developed by other researchers. For instance, by changing the raw mass ratio of D-maltose to KMnO$_4$, a series of manganese-based precursors (amorphous MnO$_2$ flower, γ-MnOOH nanorod, MnCO$_3$ cube, polyhedron, spindle, and fusiform) were synthesized and sintered at 600°C to obtain various α-Mn$_2$O$_3$ morphologies.

In contrast to the contribution of KMnO$_4$ mentioned earlier, KMnO$_4$ has also been used to oxidize low-valence manganese compounds, such as the following:

$$3\text{MnCO}_3 + 2\text{KMnO}_4 \rightarrow 5\text{MnO}_2 + \text{K}_2\text{CO}_3 + 2\text{CO}_2 \quad (5.11)$$

Using this reaction, hierarchically hollow microspheres and microcubes of MnO$_2$ were prepared by retaining the morphologies of the MnCO$_3$ precursor (Fig. 5.1). The well-defined shape of MnCO$_3$ could be adjusted by changing the synthesis conditions. The pathway to hierarchically hollow MnO$_2$ structures is shown in

Fig. 5.1G. Briefly, various morphologies of $MnCO_3$ precursors were prepared in advance via a hydrothermal method. Then, $KMnO_4$ reacted with $MnCO_3$ to form core–shell $MnCO_3@MnO_2$, in which MnO_2 was the shell and $MnCO_3$ was the core. After removing the $MnCO_3$ core by HCl etching, the MnO_2 shell retained the original framework, and hollow structures were produced. In addition, by prolonging the reaction time of $KMnO_4$ with $MnCO_3$, the thickness of MnO_2 could be increased. Similarly, Cao et al. synthesized various morphologies of Mn_2O_3, including hollow-structured spheres, cubes, ellipsoids, and dumbbells, based on morphologically controllable $MnCO_3$ precursors. Other similar approaches have been proposed to produce $LiMn_2O_4$. For instance, through the hydrothermal process, β-MnO_2 was obtained by the oxidation of $Mn(CH_3COO)_2 \cdot 4H_2O$ in the presence of peroxysulfate (($NH_4)_2S_2O_8$). Then, spinel $LiMn_2O_4$ was prepared by reacting LiOH with the as-synthesized β-MnO_2 nanorods.

Surfactants were also applied to construct metal carbonate precursors, which underwent processes similar to those of metal hydroxides. Typically, three types of highly uniform Co_3O_4 products (peanut-like, capsule-like, and rhombus topography) were synthesized from various $CoCO_3$ precursors by using $Co(CH_3COO)_2 \cdot 4H_2O$, PVP, diethylene glycol (DEG), and urea as reaction materials. A precipitation–dissolution–renucleation–growth–aggregation mechanism was proposed to explain the formation of the precursors. Initial primary precipitates were produced instantaneously when super-saturation was reached. Unstable precipitates underwent dissolution, renucleation, and crystallite growth. The $CoCO_3$ crystallization process can be described simply as follows: When heated at approximately 90°C, the urea decomposes into CO_2 and OH^- (Eq. (5.12)). Under sealed conditions, the dissolved CO_2 is largely converted to carbonate ions (Eq. (5.13)). Then, Co^{2+} combines with the generated carbonate ions to form the $CoCO_3$ precipitate (Eq. (5.14)), and finally, $CoCO_3$ is annealed in air to obtain the target Co_3O_4 (Eq. (5.15)). The formation process of $CoCO_3$ is heavily dependent on dissolution of the carbonate ion in Eq. (5.13).

$$CO(NH_2)_2 + 3H_2O \rightarrow 2NH_4^+ + CO_2 + 2OH^- \qquad (5.12)$$

$$2NH_4^+ + CO_2 + 2OH^- \rightarrow 2NH_4^+ + CO_3^{2-} + H_2O \qquad (5.13)$$

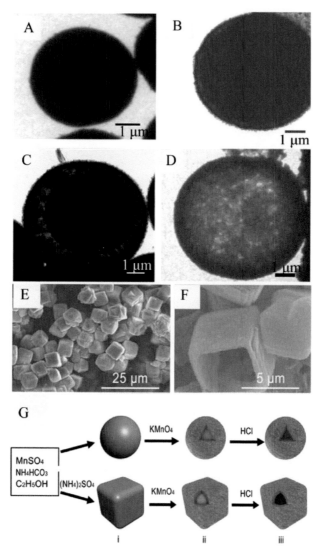

Figure 5.1 Transmission electron microscopy (TEM) images of an MnCO₃ microsphere (A), MnCO₃@MnO₂ (B) after chemical reaction (Eq. (5.11)), MnCO₃@MnO₂ (C) after partial removal of MnCO₃, and hollow MnO₂ (D) with complete removal of MnCO₃. Scanning electron microscopy (SEM) images of MnCO₃ microcubes (E) and MnO₂ microcubes (F). Schematic illustration (G) of the formation process of hierarchically hollow MnO₂ nanostructures: (i) MnCO₃ precursors with different morphologies, (ii) MnO₂@MnCO₃, and (iii) hierarchically hollow MnO₂ nanostructures. (Chen et al., 2017; Fei et al., 2008).

$$Co^{2+} + CO_3^{2-} \rightarrow CoCO_3 \quad (5.14)$$

$$6CoCO_3 + O_2 \rightarrow 2Co_3O_4 + 6CO_2 \quad (5.15)$$

Analogously, $CoCO_3$ nanostructures were successfully fabricated via a solvothermal route using PVP as a capping reagent. Intriguing anisotropic porous Co_3O_4 nanocapsules were then formed by heat treatment of $CoCO_3$ in air. In addition to the surfactant, the geometry and strength of the chelating agent play crucial roles in controlling the shape. For example, 3D hierarchically assembled lotus-shaped porous MnO_2 was synthesized by the calcination of an $MnCO_3$ precursor. The growth of lotus-like $MnCO_3$ depended on the chelating agent (citric acid) present in a simple aqueous solution. Indeed, rods, spheres, and nanoaggregates of $MnCO_3$ could also be synthesized by varying the chelating agent (citric acid, tartaric acid, and oxalic acid, respectively). Interestingly, some intriguing structures were also prepared by controlling the thermal decomposition temperature. For example, triple-shelled, cubic-like porous Mn_2O_3 was fabricated by controlling the thermal decomposition temperature of the $MnCO_3$ precursor (Fig. 5.2). The $MnCO_3$ precursor was sintered at 300°C for 1 h with a ramping rate of 1°C/min, followed by heating at 600°C for 1 h with 2°C/min. At the initial stage of calcination, a large temperature gradient (ΔT_1) existed along the radical direction, resulting in an Mn_2O_3 shell at the surface of the $MnCO_3$ core. The hierarchically porous structure was produced by two forces from opposite directions: the contraction force (F_c) from the decomposition of $MnCO_3$, which promoted the inward shrinkage of the $MnCO_3$ core, and the adhesion force (F_a) from the relatively rigid Mn_2O_3 shell, which prevented its inward contraction. With a large ΔT_1, F_c exceeded F_a during the early stage. Thus, the inner core shrank inward and became isolated from the outer shell. Similar to the first heat treatment (ΔT_1), triple-shelled Mn_2O_3 was produced by the second heat treatment (ΔT_2). With prolonged heating, F_c decreased rapidly. When F_a surpassed F_c, the direction of material movement was reversed and, as a result, the inner core contracted outward, leaving a hollow cavity in the center. Similarly, double-shelled $CoMn_2O_4$ hollow microcubes were prepared by heating $Co_{0.33}Mn_{0.67}CO_3$ precursor at 600°C for 5 h, with a ramping rate of 2°C/min in air.

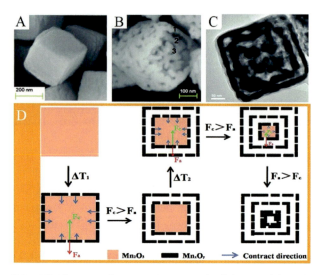

Figure 5.2 SEM images of an MnCO$_3$ nanocube (A), SEM (B), and TEM (C) images of a triple-shelled Mn$_2$O$_3$ hollow nanocube, and a schematic illustration (D) of the formation of a triple-shelled Mn$_2$O$_3$ hollow nanocube. (Chen et al., 2017; Lin et al., 2014a; Lin et al., 2014b).

Additionally, by precisely controlling the thermal treatment, Wang et al. synthesized hollow MnO$_2$ via the partial thermal decomposition of MnCO$_3$. Through a partial calcination process, the MnCO$_3$ on the surface was converted to an oxidation layer of MnO$_2$, leaving the MnCO$_3$ inner cores, and thus forming a "core@shell" (MnCO$_3$@MnO$_2$) structure. Acid treatment was applied to remove the inner cores and form porous hollow MnO$_2$. This approach was also applied to prepare other hollow MO$_x$ (M: Fe, Co, Ni, etc.). The pore sizes could also be changed by controlling the annealing temperatures. For example, Chang et al. successfully synthesized Mn$_2$O$_3$ microspheres with different pore sizes through the morphology-conserved transformation of MnCO$_3$ at different annealing temperatures. The nanoparticles within the microspheres became obviously larger, and the diameter of the microspheres decreased as the annealing temperature increased. This result might be attributable to the aggregation and regrowth of the nanoparticles, accompanied by the pores becoming larger and increasingly inhomogeneous, until disappearing completely.

5.1.3 Metal Carbonate Hydroxide Precursors

Metal carbonate hydroxides are also promising precursors for the preparation of hierarchically porous metal oxides. To synthesize a metal carbonate hydroxide, OH^- and CO_3^{2-} must exist or be generated in the reaction solution. The metal ion might form metal hydroxide, metal carbonate, or metal carbonate hydroxide, depending on its properties (e.g., the solubility of the corresponding metal salts). Therefore, not every metal ion can form a metal carbonate hydroxide. Generally, Zn-based carbonate hydroxide, Co-based carbonate hydroxide, and Ni-based carbonate hydroxide are universally used compounds.

Zn-based carbonate hydroxide is the most representative example and is thus used to introduce the metal carbonate hydroxide precursor. Various methods, such as the chemical-bath deposition method, reflux method, hydrothermal method, and solvothermal method, have been developed to synthesize zinc-based carbonate hydroxide precursors, including $Zn_5(CO_3)_2(OH)_6$ (Fig. 5.3A–E) and $Zn_4CO_3(OH)_6$ (Fig. 5.3F–J). The reaction processes can be expressed as follows:

$$5Zn^{2+} + 6OH^- + 2CO_3^{2-} \to Zn_5(CO_3)_2(OH)_6 \quad (5.16)$$

$$4Zn^{2+} + 6OH^- + CO_3^{2-} + xH_2O \to Zn_4(CO_3)(OH)_6 \cdot xH_2O \quad (5.17)$$
$$(x \text{ may be 0 or 1})$$

For example, in Jing's work, a plate-like $Zn_5(CO_3)_2(OH)_6$ precursor with edge thicknesses of approximately 19 nm was prepared in the presence of urea, and then annealed at 400°C for 2 h to generate porous ZnO nanoplates. Additionally, a 3D porous ZnO architecture consisting of interconnected nanosheets was fabricated by the conversion of a layered $Zn_4(CO_3)(OH)_6 \cdot H_2O$ precursor. In the aforementioned preparation, urea played an important role in determining the final morphology. Similarly, assisted by urea, Lei et al. fabricated porous ZnO microspheres by calcining $Zn_4(CO_3)(OH)_6$ microspheres. During the annealing process, a porous structure was probably formed because of the release of H_2O and CO_2 in the thermal decomposition of the precursors. The thermal decomposition could be described as follows:

$$Zn_x(CO_3)_y(OH)_6 \cdot zH_2O \to xZnO + yCO_2 \uparrow + (z+3)H_2O \uparrow \quad (5.18)$$

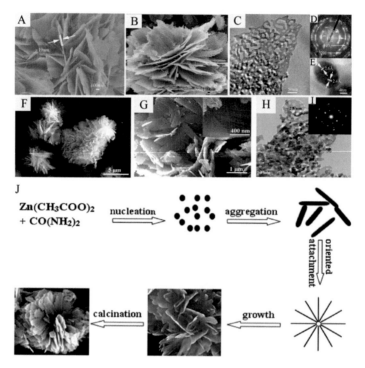

Figure 5.3 SEM image of the $Zn_5(CO_3)_2(OH)_6$ precursor (A), which is composed of plate-like nanostructures with edge thicknesses of approximately 19 nm, and porous ZnO nanoplates (B). Typical TEM image of an individual porous ZnO nanoplate (C). Selected-area electron diffraction (SAED) patterns (D) and high-resolution TEM (HRTEM) image (E) of porous ZnO nanoplates. Reproduced with permission. Copyright 2008, Wiley-VCH. SEM images of the $Zn_4CO_3(OH)_6$ precursor (F) and 3D porous architecture of ZnO (G). Magnified TEM image of the ZnO nanosheets (H) and corresponding SAED pattern of H (I). Schematic illustration of the formation of porous structures (J). (Chen et al., 2017; Li et al., 2010).

The Co-based intermediate compound is another important candidate precursor for the synthesis of Co_3O_4 or CoO. For example, Xiong et al. synthesized chrysanthemum-like $Co(CO_3)_{0.5}(OH) \cdot 0.11H_2O$ through a hydrothermal route with the assistance of urea and sodium chloride. The precursor was then transformed into mesoporous Co_3O_4 under normal annealing conditions. Cubic CoO nanonets were successfully prepared by annealing a monoclinic $Co_2(OH)_2CO_3$-nanosheet precursor in an Ar atmosphere. Similarly, Co_3O_4 nanorods, nanobelts, nanosheets, and cubic/octahedral nanoparticles exposing the high-energy (110) crystal plane were successfully synthesized by annealing the Co(CO$_3$)$_{0.5}$(OH)·0.11H$_2$O precursor at 250°C. Moreover, cobalt–nickel bimetallic hydroxide carbonate and cobalt–manganese bimetallic hydroxide carbonate with hierarchical structures were synthesized as precursors to obtain the final hierarchically porous products. Specifically, a sea-urchin-like NiCo$_2$O$_4$ spinel structure was prepared through a calcination process with morphology conserved from a bimetallic (Ni, Co) carbonate hydroxide. The bimetallic carbonate hydroxide precursor was formed via a sequential crystallization process. Specifically, monometallic nickel carbonate hydroxide first nucleated, then evolved into flower-like microspheres, grew into bimetallic hydroxide carbonate nanorods via localized dissolution recrystallization, and finally formed a sea-urchin-like structure. Additionally, nickel–cobalt-oxide microspheres consisting of mesoporous thorn arrays were obtained by the calcination of a well-designed NiCo(OH)$_2$CO$_3$.

Weak acid salts combined with (NH$_4$)$_2$CO$_3$ have also been used to provide CO$_3^{2-}$ and OH$^-$. For instance, 3D nest-like porous ZnO was synthesized by Wang et al. by annealing the Zn$_5$(CO$_3$)$_2$(OH)$_6$ precursor. The precursor was obtained through a hydrothermal process in which OH$^-$ and CO$_3^{2-}$ were provided by the hydrolysis of zinc acetate and the dissociation of (NH$_4$)$_2$CO$_3$, respectively.

Hexamethylenetetramine is also useful for the preparation of metal carbonate hydroxide precursors. When the reaction temperature exceeds 120°C, hexamethylenetetramine decomposes, releasing OH$^-$ and CO$_3^{2-}$ in water. For example, Ni$_2$(OH)$_2$CO$_3$ microspheres assembled from nanosheets were obtained through a hydrothermal route in the presence of hexamethylenetetramine. Hierarchically porous NiO microspheres were fabricated by further heat treatment at 300°C for 2 h.

As mentioned earlier, employing surfactants, such as SDS, Pluronic F127, and cetyltrimethylammonium bromide (CTAB), is an effective approach to achieving metal carbonate hydroxide precursors with various morphologies. For instance, urchin-like hollow Co$_3$O$_4$ spheres were successfully synthesized by Chen et al. via the thermal decomposition of a Co(CO$_3$)$_{0.5}$(OH)·0.11H$_2$O precursor obtained with the assistance of CTAB.

5.1.4 Metal–Organic Framework Precursors

Metal–organic frameworks (MOFs) are also among the most promising candidate precursors for hierarchically porous materials. As the linkers of MOFs, ligands are of great importance because they often contain functional groups (e.g., –OH, –COOH, and –NH$_2$) that provide coordination sites. Based on the different styles of ligands, we divide the precursors of MOFs into several classes: metal carboxylate complexes, metal hydroxide acetate complexes, metal alkoxide complexes, metal glycolate complexes, Prussian blue (PB) coordination complexes, and metal imidazolyl/pyridyl complexes.

5.1.5 Other Precursors

In addition to the precursors mentioned earlier, some metal inorganic salts, such as metal hydroxyoxides, hydrotalcite-like salts, and nitrates, are occasionally used as solid precursors to synthesize their nanostructured metal oxides after chemical/thermal conversion.

5.2 Application in LIBs

The applications of hierarchically porous structured materials in LIBs generally fall into two broad categories: anode materials and cathode materials.

5.2.1 Anode Materials

Transition-metal-based oxides, such as monometal oxides (e.g., cobalt oxide, nickel oxide, iron oxide, and manganese oxide) and mixed-metal oxides ($A_xB_{3-x}O_4$; A, B = Co, Ni, Zn, Mn, Fe, etc.), have been employed as anode materials because of their relatively high theoretical capacities (e.g., 890 mAh/g for Co_3O_4, 718 mAh/g for NiO, 1007 mAh/g for Fe_2O_3, and 1019 mAh/g for Mn_2O_3) compared to commercialized graphite ($\approx$372 mAh/g). These high capacities are attributable to the conversion reaction between lithium and metal oxides in the electrolyte, as follows (Eq. (5.19)):

$$M_xO_y + 2\gamma Li^+ + 2\gamma e^- \rightarrow xM + 2\gamma Li_2O \qquad (5.19)$$

Despite the high capacity, the practical use of metal oxides as anode materials remains limited by the high irreversible capacity losses that occur during the first charge/discharge cycle and deterioration of the active materials during long cycling. The former problem is often ascribed to the innate characteristics of most anode materials, whereas the latter results from large volume changes (e.g., ≈100% volume expansion for Co_3O_4 and Fe_3O_4) during the lithium insertion/extraction process.

Cobalt oxides with hierarchically porous structures can be easily formed via a morphology-conserved transformation method and exhibit improved lithium-ion storage performances. There are two main types of cobalt oxides: CoO and Co_3O_4. Their theoretical capacities are 715 and 890 mAh/g, respectively. Voltage plateaus for the CoO electrodes are observed at 0.7 and 2.1 V during the initial discharge and charge, respectively. In the case of the Co_3O_4 electrodes, the voltage plateaus occur at ≈0.8–1.3 V during initial discharge, and 2.2 V during the charge process. Substantial attention has been focused on Co_3O_4 materials because of their higher capacities and easy preparation. For instance, hierarchically hollow Co_3O_4 nanoneedles originating from β-$Co(OH)_2$ nanoneedles were well designed to improve the electrochemical performance of LIBs. The results of electrochemical testing revealed that the first discharge capacity of the as-synthesized sample was approximately 1290 mAh/g at 150 mA/g between 3 V and 10 mV and that a reversible capacity of 1079 mAh/g was maintained after 50 cycles. This good electrochemical performance was attributed to the short transport distance of the lithium ions and the enhanced interconnection among individual nanoparticles. A porous Co_3O_4 nanobelt array fabricated from a uniform and well-aligned $Co(CO_3)_{0.5}(OH)_{0.11}H_2O$ nanobelt array precursor was found to exhibit good electrochemical performance. Charge/discharge testing revealed that the Co_3O_4 nanobelt array retained a specific capacity of 770 mAh/g over 25 cycles at 177 mA/g. Even at high current densities of 1670 and 3350 mA/g, specific capacities of 510 and 330 mAh/g, respectively, were achieved after 30 cycles. Single- (S-Co), double- (D-Co), and triple-shelled (T-Co) hollow spheres (Fig. 5.4) assembled from Co_3O_4

nanosheets were successfully synthesized by calcining the cobalt glycolate precursor. Based on initial charge/discharge testing, the discharge capacities of S-Co, D-Co, and T-Co were approximately 1199.3, 1013.1, and 1528.9 mAh/g, respectively. According to the cycling performance testing results, S-Co, D-Co, and T-Co maintained capacities as high as 680, 866, and 611 mAh/g, respectively, after 50 cycles at a rate of C/5; all these values were better than that of a commercial sample (C-Co). Furthermore, even at a high current rate of 2 C, D-Co was able to deliver a capacity of 500.8 mAh/g, indicating that D-Co has good rate capability.

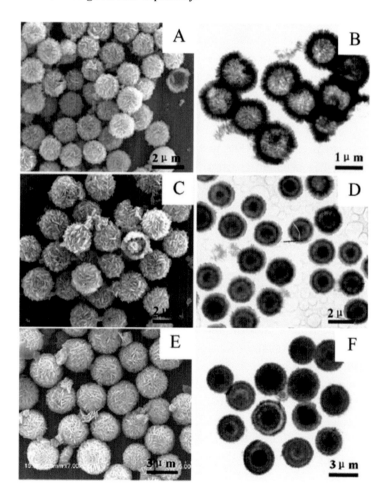

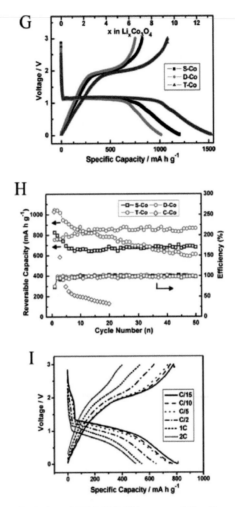

Figure 5.4 SEM (A,C,E) and TEM (B,D,F) images of the three samples: (A,B) S-Co, (C,D) D-Co, and (E,F) T-Co. (G) The first cycle discharge/charge curves of S-Co, D-Co, and T-Co. (H) Cycling performance of the three as-prepared samples and commercial Co_3O_4 product (C-Co) at a current rate of C/5 (178 mA/g). (I) Charge/discharge curves of D-Co at different current densities. (Chen et al., 2017; Wang et al., 2010).

Manganese-based oxides have also been extensively researched because of their low cost, environmental friendliness, high capacity, and low reaction voltage (0.2–0.5 V during initial discharge). A variety of morphologies and crystallographic structures have been obtained from the conversion of different types of precursors.

The crystallographic structures of manganese oxide mainly include cubic rock salt (MnO, 756 mAh/g), inverse spinel (Mn$_3$O$_4$, 937 mAh/g), hexagonal corundum (Mn$_2$O$_3$, 1019 mAh/g), and a manganese dioxide structure (MnO$_2$, 1223 mAh/g). In these reports, manganese-based oxides with hierarchically porous structures were found to exhibit enhanced lithium storage capacities. Mn$_2$O$_3$, which is a well-known functional transition-metal oxide with structural flexibility, has attracted extensive attention because of its distinctive physicochemical properties. For example, porous Mn$_2$O$_3$ nanomaterials were fabricated by Qian et al. via the conversion of a simple Mn(OH)$_2$ precursor. The as-prepared Mn$_2$O$_3$ exhibited a high and stable reversible capacity. Specifically, the porous Mn$_2$O$_3$ nanoflowers could maintain a capacity of ≈521 mAh/g after 100 cycles at a current density of 300 mA/g, indicating that the porous structure was crucial for the enhanced electrochemical performance of Mn$_2$O$_3$. Triple-shelled Mn$_2$O$_3$ hollow nanocubes derived from MnCO$_3$ also exhibited excellent electrochemical performance. When evaluated as an anode material for LIBs, this material could deliver reversible capacities of 606 and 350 mAh/g at current densities of 500 and 2000 mA/g, respectively.

Iron oxides have also been investigated intensively as promising anode materials for LIBs because of their high theoretical capacity, low cost, environmental friendliness, and high resistance to corrosion. Two typical iron oxide phases, hematite (α-Fe$_2$O$_3$, 1007 mAh/g) and magnetite (Fe$_3$O$_4$, 924 mAh/g), could be obtained via the morphology-conserved transformation method. The discharge plateaus for iron oxides lie at approximately 0.8 V. Fe$_2$O$_3$ nanostructures with various features have been evaluated to improve the resulting electrochemical properties. For instance, Fe$_2$O$_3$ microboxes with a well-defined hollow structure and hierarchical shell originating from a PB precursor were found to display a high specific capacity and excellent cycling performance, with the highest reversible capacity of 945 mAh/g in the 30th cycle at 200 mA/g. As described above, spindle-like porous α-Fe$_2$O$_3$ has shown enormously enhanced lithium storage performance. The capacity of the porous α-Fe$_2$O$_3$ was 911 mAh/g after 50 cycles at a rate of 0.2 C. Even when cycled at a high rate (10 C), a reversible capacity of 424 mAh/g could be obtained.

Copper oxides Cu$_2$O (375 mAh/g) and CuO (674 mAh/g) have also been utilized as anode materials because they are nontoxic and naturally abundant, and can be prepared by a simple method. For example, the as-prepared bundle-like CuO mentioned earlier exhibited excellent electrochemical performance, with a high rate capability. The initial discharge capacity of CuO was 1179 mAh/g at a rate of 0.3 C, and a capacity of 666 mAh/g was retained after 50 cycles. Even at a high rate (6 C), a capacity of 361 mAh/g was maintained.

Binary transition-metal oxides have also been widely studied as anode materials for LIBs because of their higher theoretical specific capacity, superior rate performance, and better cycling stability compared to mono-transition-metal oxides. Obtaining these mixed-metal oxides via the thermal treatment of diverse precursors is simple. The electrical/ionic conductivity of Ni–Co-based oxides was greatly improved, leading to enhanced electrochemical properties, especially the rate capability. For instance, mesoporous $Ni_{0.37}Co$ oxide nanoprisms with a yolk–shell structure derived from Ni–Co-based hydroxide acetate could deliver an initial discharge capacity of 1394.4 mAh/g and maintain a reversible capacity of ≈1028.5 mAh/g, after 30 cycles at a current density of 200 mA/g. Among these Ni–Co-based oxides, $NiCo_2O_4$, with a spinel structure (i.e., in which nickel occupies the octahedral sites, and cobalt is distributed over both octahedral and tetrahedral sites), has attracted the most attention because of its significantly enhanced electrochemical performance. For instance, mesoporous $NiCo_2O_4$ microspheres originating from $Ni_{0.33}Co_{0.67}CO_3$ were shown to retain a reversible capacity of 1198 mAh/g after 30 cycles, at a current density of 200 mA/g. Even at a current density of 800 mA/g, a capacity of 705 mAh/g was maintained after 500 cycles. Other binary metal oxides, such as $CoMn_2O_4$, have higher electronic conductivities than MnO_x. Double-shelled hollow $CoMn_2O_4$ microcubes derived from $Co_{0.33}Mn_{0.67}CO_3$ cubes were observed to have a discharge capacity of ≈830 mAh/g at a current density of 200 mA/g, and maintained a capacity of 624 mAh/g after 50 cycles. Foam-like porous spinel $Mn_xCo_{3-x}O_4$ was obtained from $Mn_3[Co(CN)_6]_2 \cdot nH_2O$ nanocubes. Electrochemical testing revealed that the initial discharge capacity was 1395 mAh/g at a current density of 200 mA/g, and that a high charge capacity of 733 mAh/g was retained after 30 cycles.

Anode materials with lower charge voltages could deliver higher energy densities. $ZnMn_2O_4$ has attracted extensive attention because its operating voltage is much lower than those of Co- or Fe-based oxides and because it is low cost and environmentally friendly. For example, $ZnMn_2O_4$ ball-in-ball hollow microspheres obtained from ZnMn-glycolate (Fig. 5.5) were found to deliver an initial discharge/charge capacity of ≈945/662 mAh/g at a current density of 400 mA/g. During the cycling performance testing, the discharge capacity gradually decreased to 490 mAh/g after approximately 50 cycles, maintained a steady state for dozens of cycles and then began to increase, becoming as high as 750 mAh/g after 120 cycles. Correspondingly, the Coulombic efficiency was approximately 70% for the first cycle, quickly increased to 98% after several cycles, and then stabilized at nearly 100% for the remaining cycles. After cycling for 120 cycles at a current density of 400 mA/g, the same cell was subjected to rate capability testing. This cell was able to deliver specific capacities of 683, 618, 480, and 396 mAh/g at current densities of 600, 800, 1000, and 1200 mA/g, respectively. The enhanced electrochemical performance of the $ZnMn_2O_4$ ball-in-ball hollow microspheres is attributable to the following: The small average size of the primary nanoparticles could reduce the diffusion distance of Li^+, and the void space in the porous structure could serve as an electrolyte reservoir, leading to good rate capability. More importantly, the unique ball-in-ball hollow morphology of $ZnMn_2O_4$ could significantly improve the structural integrity, by alleviating the mechanical strain induced by volume expansion during repeating cycling to some extent, resulting in excellent cycling stability.

Metal vanadates can also be used as anode materials. Indeed, the design and synthesis of metal vanadates with various morphologies and structures have attracted tremendous interest. For instance, porous $Co_2V_2O_7$ hexagonal nanoplatelets were obtained via thermal treatment of a $Co_2V_2O_7 \cdot nH_2O$ precursor and exhibited good cycling stability and rate capability. Specifically, porous $Co_2V_2O_7$ maintained a reversible capacity of 866 mAh/g with almost 100% capacity retention after 150 cycles at a current density of 0.5 A/g, and delivered specific capacities of 813 mAh/g, 666 mAh/g, 594 mAh/g, 518 mAh/g, and 344 mAh/g at 0.2 A/g, 0.5 A/g, 1 A/g, 2 A/g, and 5 A/g, respectively. Enhanced lithium storage performance has also been reported for many other metal oxides, such as NiO, $MnCo_2O_4$, $Mn_{1.5}Co_{1.5}O_4$, $CoFe_2O_4$, $ZnCo_2O_4$, and $CaSnO_3$, which can also be produced by the morphology-conserved transformation method.

Application in LIBs | 137

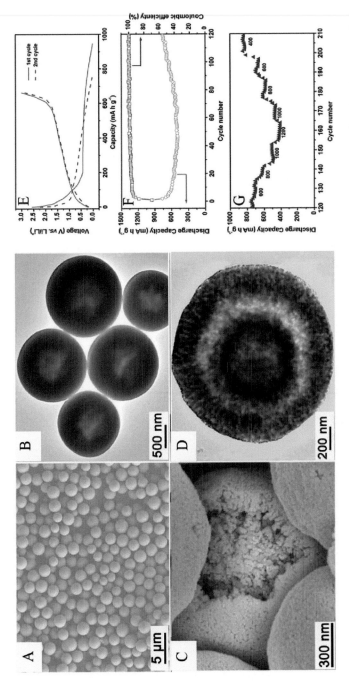

Figure 5.5 Typical field-emission SEM (FESEM) (A, C) and TEM (B, D) images of the ZnMn–glycolate precursor (A, B) and ZnMn$_2$O$_4$ ball-in-ball hollow microspheres (C, D). The first and second charge/discharge profiles (E) obtained at a current density of 400 mA/g. Cycling performance and the corresponding Coulombic efficiency (F) at 400 mA/g. Rate capabilities (G) at various current densities. (Chen et al., 2017; Zhang et al., 2012).

5.2.2 Cathode Materials

The currently available cathode materials include layered LiCoO$_2$ (≈140 mAh/g), spinel LiMn$_2$O$_4$ (≈120 mAh/g), and olivine LiFePO$_4$ (≈170 mAh/g). To meet the demands for higher energy and power densities in LIBs, new structured cathode materials with higher specific capacities and high voltages must be developed. Among the metal-based cathode materials reported to date, lithium-rich layered oxide materials (i.e., xLi$_2$MnO$_3$·(1−x)LiMO$_2$ [M = Mn, Ni, Co, Fe, Cr, etc.]) have attracted much attention because of their high capacities (>250 mAh/g) and high operation voltages (>4.6 V at room temperature). For instance, Wu et al. produced hierarchical nanostructured Li$_{1.2}$Ni$_{0.2}$Mn$_{0.6}$O$_2$ with exposed (010) planes (HSLR), which was converted from a quasi-spherical Ni$_{0.2}$Mn$_{0.6}$(OH)$_{1.6}$ precursor (Fig. 5.6). High specific discharge capacities of 230.8, 216.5, 188.2, 163.2, and 141.7 mAh/g were obtained at rates of 1 C, 2 C, 5 C, 10 C, and 20 C at 2–4.8 V, respectively. This outstanding high rate capability could be ascribed to two key factors: the unique hierarchical structure, including the special directional alignment of the nanoplates that facilitated the rapid diffusion of Li$^+$, and the hierarchical structure that provided efficient 3D electron transport networks (Yu et al., 2015).

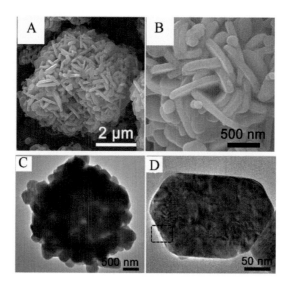

Application in LIBs | 139

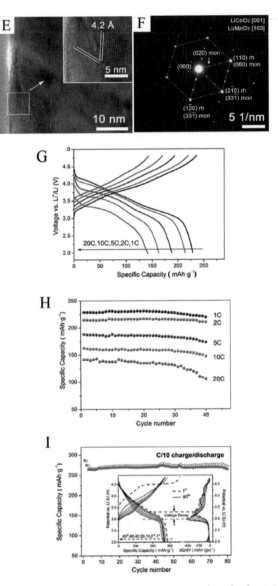

Figure 5.6 FESEM images (A, B) and TEM images (C, D) of HSLR. HRTEM image (E) of the region marked in (D). The inset is a magnified image of the frame shown in (E). (F) Corresponding SAED pattern of D. (G) Charge/discharge curves of HSLR at various rates. (H) Cycling performance of HSLR at different rates. (I) Cycling performance of HSLR materials at C/10; the inset shows the corresponding voltage profile and dQ/dV plots. (Chen et al., 2017; Chen et al., 2014).

$LiNi_{1/3}Co_{1/3}Mn_{1/3}O_2$ is another typical layered cathode material in which the valences of the nickel, cobalt, and manganese ions are +2, +3, and +4, respectively. This material exhibits an enhanced electrochemical performance relative to $LiCoO_2$, $LiNiO_2$, and $LiMnO_2$. $LiNi_{1/3}Co_{1/3}Mn_{1/3}O_2$ is typically obtained through first synthesizing metal carbonate precursors and then reacting them with lithium sources at high temperature. Hierarchically porous structured $LiNi_{1/3}Co_{1/3}Mn_{1/3}O_2$ can achieve remarkable performances in terms of high rate capabilities and long cycling stabilities. For example, $LiNi_{1/3}Co_{1/3}Mn_{1/3}O_2$ hollow microspheres exhibited a high discharge capacity of 157.3 mAh/g at 0.2 C after 100 cycles, and 120.5 mAh/g at 0.5 C after 200 cycles. Even at a high rate (5 C), a high capacity of 114.2 mAh/g was maintained. Spinel $LiMn_2O_4$ and $LiNi_{0.5}Mn_{1.5}O_4$ possess 3D lithium-ion channels and thus exhibit good rate capabilities. For instance, porous $LiMn_2O_4$ spheres exhibited stable cycling ability and high rate capability. The capacity retention was 94% after 100 cycles at 1 C. Additionally, the discharge capacity remained at ≈83 mAh/g at 20 C. The spinel $LiMn_2O_4$ nanorods synthesized by Cui et al. delivered high charge storage capacities at high power rates. Cycling stability tests revealed that capacity retention of more than 85% was maintained for over 100 cycles. Porous cubic $LiMn_2O_4$ (Fig. 5.7) derived from an $MnCO_3$ precursor exhibited an excellent high rate capability and a long-term cycle life. At a high rate of 30 C, this material could still deliver a reversible capacity of 108 mAh/g and, more importantly, its capacity retention was 80% at 10 C after 4000 cycles. Additionally, $LiNi_{0.5}Mn_{1.5}O_4$ hollow microspheres/microcubes consisting of nanosized particles showed high capacity, excellent cycling stability, and exceptional rate capability. These structures could deliver a discharge capacity of approximately 120 mAh/g at 0.1 C. Even at 20 C, the retained capacity was still 104 mAh/g. Furthermore, the capacity retention could reach 96.6% at 2 C after 200 cycles. $LiNi_{0.5}Mn_{1.5}O_4$ porous nanorods with an ordered $P4_332$ phase derived from an MnC_2O_4 precursor could deliver reversible capacities of 140 and 109 mAh/g at rates of 1 and 20 C, respectively. Impressively, a capacity retention of 91% was maintained after 500 cycles at 5 C. This remarkable performance was attributed to the porous 1D nanostructures, which could accommodate strain relaxation during cycling and provide short lithium-ion-diffusion distances along the confined dimension.

Application in LIBs | 141

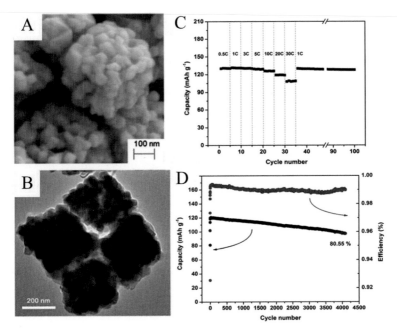

Figure 5.7 SEM image (A) and TEM image (B) of porous LiMn$_2$O$_4$. Rate performance of the as-prepared porous LiMn$_2$O$_4$ (C). Deep cycle performance at a rate of 10 C for porous LiMn$_2$O$_4$ (D). (Chen et al., 2017; Lin et al., 2014a; Lin et al., 2014b).

Other cathode materials, such as V$_2$O$_5$, derived from various precursors have also been investigated. For instance, Lou et al. reported 3D porous V$_2$O$_5$ hierarchical microspheres with good cycling ability and excellent rate capability. Indeed, a stable capacity of 130 mAh/g at 0.5 C was maintained after 100 cycles. Even at 30 C, a capacity of 105 mAh/g was achieved. Notably, several suitable strategies, including coating with carbonaceous materials and TiO$_2$, doping a spinel phase on the surface of porous materials, or using functional additives, are often applied to generate hierarchically porous micro-/nanomaterials with improved electrochemical performance. For example, Wang et al. synthesized carbon-coated α-Fe$_2$O$_3$ hollow nanohorns onto carbon nanotube (CNT) backbones to enhance electron transport and prevent agglomeration. This unique hybrid structure exhibited a stable capacity of 800 mAh/g over 100 cycles at a current density of 500 mA/g. Even at high current densities of 1000–3000 mA/g, capacities of 420–500 mAh/g were maintained. Cubic Mn_2O_3@TiO_2 established on Mn_2O_3 porous nanocubes displayed superior charge/discharge performance. A reversible capacity of 263 mAh/g at 6000 mA/g was achieved for Mn_2O_3@TiO_2, whereas unmodified Mn_2O_3 exhibited a value of only 9.7 mAh/g. Moreover, using additives in electrolytes is one of the most effective approaches to enhance the electrochemical performance of electrode materials because of the in situ formation of a uniform interphase film that isolates direct contact between the electrolyte and electrode and prevents metal ion dissolution. Therefore, a hybrid approach must be developed to further improve the electrochemical performances of these hierarchically porous micro-/nanostructure materials.

5.3 Conclusion

In conclusion, several representative precursors, such as metal hydroxides, metal carbonates, metal hydroxide carbonates, and MOFs, have been used to create hierarchically porous micro-/nanostructured materials via the morphology-conserved transformation approach.

Hierarchically porous micro-/nanostructured materials have the unique feature of micro- or sub-microsized architectures consisting of nanosized units and are endowed with the advantages of both hierarchical micro-/nanostructures and porous structures, such as high specific surface areas, porosity, and short ion-/electron-diffusion pathways. When utilized in LIBs, these materials can avoid agglomeration during electrochemical cycling and minimize interfacial contact resistance. Nanosized units can enhance Li diffusion and alleviate inner stress, thereby improving the electrode capacity and rate capability. Furthermore, the porosity of the materials can enhance the contact surface areas between the electrode materials and electrolytes, and thus facilitate Li^+ diffusion. Last, porosity can also alleviate the strain induced by volume changes during lithium insertion/extraction, leading to good long-term cycling performance.

Questions

1. Please describe hierarchically porous micro-/nanostructures.
2. How would you classify the precursor preparation techniques for hierarchical porous micro/nanostructures?
3. What is the mechanism of precursor preparation techniques for hierarchical porous micro/nanostructures?
4. What are the advantages and disadvantages of hierarchical porous micro/nanostructure for positive and negative electrodes of lithium-ion batteries?

References

Chen L., Su Y., Chen S., Li N., Bao L., Li W., Wang Z., Wang M., and Wu F. (2014). *Adv. Mater.*, **26**, 6756.

Chen M., Zhang Y. G., Xing L. D., Qiu Y. C., Yang S. H., and Li W. S. (2017). *Adv. Mater.*, **29**(48), 1607015.

Fei J. B., Cui Y., Yan X. H., Qi W., Yang Y., Wang K. W., He Q., and Li J. B. (2008). *Adv. Mater.*, **20**, 452.

Jiang X., Wang Y., Herricks T., and Xia Y. (2004). *J. Mater. Chem.*, **14**, 695.

Li J., Fan H., and Jia X. (2010). *J. Phys. Chem. C*, **114**, 14684.

Lin H. B., Hu J. N., Rong H. B., Zhang Y. M., Mai S. W., Xing L. D., Xu M. Q., Li X. P., and Li W. S. (2014a). *J. Mater. Chem. A*, **2**, 9272.

Lin H. B., Rong H. B., Huang W. Z., Liao Y. H., Xing L. D., Xu M. Q., Li X. P., and Li W. S. (2014b). *J. Mater. Chem. A*, **2**, 14189.

Wang X., Wu X. L., Guo Y.-G., Zhong Y., Cao X., Ma Y., and Yao J. (2010). *Adv. Funct. Mater.*, **20**, 1680.

Yu L., Guan B., Xiao W., and Lou X. W. (2015). *Adv. Energy Mater.*, **5**, 1500981.

Zhang G., Yu L., Wu H. B., Hoster H. E., and Lou X. W. (2012). *Adv. Mater.*, **24**, 4609.

Chapter 6

Application in Perovskite Solar Cells

Organic–inorganic hybrid perovskite solar cells (OIH-PSCs) have attracted keen attention because of their rapid increase in power conversion efficiency (PCE) since 2009. The highest reported PCE is approaching that of monocrystalline silicon solar cells. It is unprecedented for a new kind of solar cells to show such a comparable performance to traditional commercial solar cells in such a short time. It all has to do with the exceptional hybrid perovskite materials (ABX_3: A = CH_3NH_3, $HC(NH_2)_2$; B = Pb, Sn; X = Cl, Br, I), which have demonstrated promising optoelectronic properties, including high absorption coefficient, high mobility, long balanced carrier diffusion lengths, low exciton binding energy, and large polarons.

In spite of their impressively high PCEs, the OIH-PSCs commonly suffer from poor stability, especially thermal instability due to the presence of organic ions, which could be decomposed at high humidity and temperature. To overcome this problem, replacing the organic ions with inorganic ions (such as Cs^+, Rb^+, etc.) to form pure inorganic perovskites (I-PVKs) should be a foreseeable strategy. Encouragingly, many previously prepared inorganic PVKs ($CsSnI_3$, $CsPbI_3$, $CsPbBr_3$, etc.) that could be easily synthesized by solution processes have been proved to have similar optoelectronic properties with their organic counterparts. Promisingly, these I-PVKs have shown obviously improved stability (mainly including light,

---

Materials and Interfaces for Clean Energy
Shihe Yang and Yongfu Qiu
Copyright © 2022 Jenny Stanford Publishing Pte. Ltd.
ISBN 978-981-4877-66-4 (Hardcover), 978-1-003-14223-2 (eBook)
www.jennystanford.com

thermal, and electron beam stability) compared with their organic counterparts. Due to the promising stability and optoelectronic properties, I-PVKs have been more and more studied as alternative light absorbers in PSCs and the PCEs have been increasing gradually in the last few years. The first demonstration on inorganic perovskite solar cells (I-PSCs) that used a Schottky junction structure based on CsSnI$_3$ was reported in 2012, which only exhibited a PCE of 0.88%. The I-PVKs based on the common device structure of PSCs were first constructed in 2014, which also used CsSnI$_3$ as light absorber and obtained a PCE of 2.02%. Soon after, more stable CsPbX$_3$ (X = Cl, Br, and I) PVKs have been exploited in PSCs, which boosted the PCE to over 10% in a short time, and until now, a PCE as high as 13.21% has been achieved by Bi-doped CsPbI$_3$. Meanwhile, many other I-PVKs have been theoretically predicted and synthesized for PSCs. Therefore, tremendous progresses have been made on both materials and devices for I-PSCs in the last few years.

In this chapter, we focus on the I-PVKs by introducing some application examples of materials and interfaces in solar cells.

6.1 Working Principle and Characterization of Solar Cell

At interfaces between n-type and p-type semiconductors, there is often a built-in electric field, and at interfaces between donor and acceptor molecules, there is often a driving force for electron transfer. So when light is incident, the generated holes and electrons will drift in the opposite directions under the action of the built-in electric field or the driving force for electron transfer. The holes will move in the positive direction, and the electrons will move in the negative direction, thus forming a potential difference between the two electrodes and then a photocurrent. As shown in Fig. 6.1, the electron selection layer is a material with a higher transmission rate for electrons and a lower transmission rate for holes. Such material is generally called an n-type semiconductor material. Similarly, the hole selection layer is generally a p-type semiconductor material with a higher transmission rate for holes. From the band point of view, the energy barrier of electron injection into the hole transport layer is higher, thus blocking the electron. In the same way, the barrier of hole injection into electron transport layer is also higher. The energy level gradient and built-in electric field formed between each material inside the device ensure that electrons and holes enter the external circuit along their respective transmission paths.

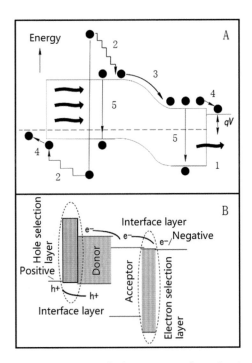

Figure 6.1 Structure diagram of photogenerated carrier band model in semiconductor (A) and solar cell device (B).

The performance of solar cells is usually characterized by the J-V curve (Fig. 6.2). A J-V curve is obtained by applying a continuously varying bias voltage to the solar device under simulated sunlight irradiation and by measuring the corresponding current density. Several important parameters related to energy conversion efficiency (PCE), including open circuit voltage (V_{oc}), short circuit current (J_{sc}), filling factor (FF), incident light power (P_{in}), and maximum power point (P_{max}), can be obtained from the curve. The relationships between them are expressed as follows:

$$FF = P_{max}/J_{sc} \cdot V_{oc}$$
$$PCE = P_{max}/P_{in} = FF \cdot J_{sc} \cdot (V_{oc}/P_{in}) \cdot 100\%$$

J_{sc} is closely related to the bandgap of the material. The smaller the bandgap, the absorption spectrum of the material can cover the solar spectrum to a greater extent, resulting in more photon conversion to electric energy. V_{oc} is largely determined by the energy gap between n and p materials. FF reflects the ideal degree of the device diode, which is affected by the series-parallel resistance of the device. P_{max} is the maximum value of the power in the fourth quadrant, which represents the actual working state of the battery. And its ratio to the incident light power is the energy conversion efficiency of the device, which is essentially the most concerned parameter of a solar cell.

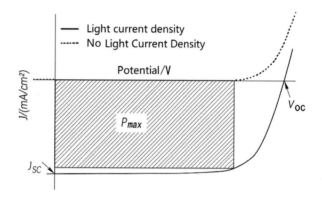

Figure 6.2 Typical J-V curves of solar cells.

6.2 Crystal Structures of I-PVKs

The basic crystal structure of I-PVKs has the form of ABX_3 (1-1-3), which includes the typical compounds of $CsSnI_3$, $CsSnBr_3$, $CsPbI_3$, $CsPbBr_3$, etc. However, some derivative structures have been exploited, such as A_2BX_6 (2-1-6), $A_3B_2X_9$ (3-2-9), and $A_2B^{1+}B^{3+}X_6$ (2-1-1-6), as illustrated in Fig. 6.3. The A_2BX_6 structure could be seen as being derived from ABX_3 PVK by removing half of the B-site cations in a checkerboard pattern, in which B-site cation in this structure is at +4 oxidation state to meet the charge neutrality requirement. The typical compounds include Cs_2SnI_6, Cs_2PbBr_6, Cs_2PbI_6, etc. The

$A_3B_2X_9$ structure could be imagined as being derived from ABX_3 PVK by removing one in three B-site cations at the top, including the typical compounds of $Cs_3Sb_2I_9$, $Rb_3Sb_2I_9$, $Cs_3Bi_2I_9$, etc., and in order to maintain charge neutrality, the B-site cation is at the +3 oxidation state. The $A_2B^{1+}B^{3+}X_6$ structure could be understood as cubic perovskite with a double unit cell, in which every pair of adjacent B^{2+} cations at the top are replaced by one B^{1+} and one B^{3+} cation. The typical compounds include $Cs_2BiAgCl_6$, $Cs_2BiAgBr_6$, etc.

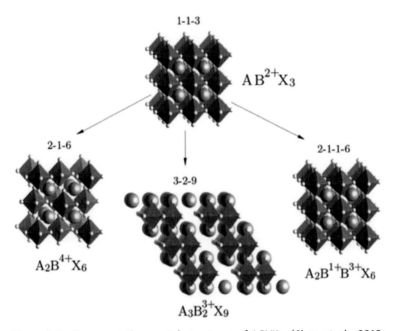

Figure 6.3 Representative crystal structures of I-PVKs. (Chen et al., 2018; Giustino and Snaith, 2016).

6.3 Lead-Based Inorganic Perovskites

6.3.1 CsPbI₃ Perovskite Solar Cells

$CsPbI_3$ is an I-PVK formed by substituting the organic cation in $MAPbI_3$ with cesium normally. $CsPbI_3$ has rarely been reported as a light absorber in PSCs because it usually exhibits a yellow non-perovskite phase at room temperature, while the desired black cubic PVK phase with a bandgap of 1.73 eV is not stable at room temperature in ambient conditions. Eperon et al., for the first time, applied black α-CsPbI$_3$ in PSCs because they found this black phase could be stable in its black phase when it was processed in an air-free atmosphere. Furthermore, they found that by adding a small amount of HI in the CsPbI$_3$ precursor solution, the transition temperature of yellow phase to black phase could be lowered to as low as 100°C, which enabled the device processing at low temperature (as illustrated in Fig. 6.4a,b). The role of HI in obtaining stable black phase at low temperature was interpreted as the formation of smaller crystals, which tended to induce phase transition at low temperature due to the generation of lattice strain. The black α-CsPbI$_3$ has been applied in PSCs based on three different device structures, including planar regular, mesoporous, and inverted planar (Fig. 6.4c), which exhibited the PCEs of 2.9, 1.3, and 1.7%, respectively (Fig. 6.4d). The more efficient planar PSCs than mesoporous PSCs implied that both electrons and holes might have a significant diffusion length in α-CsPbI$_3$, and the carrier transport in the α-CsPbI$_3$ was likely superior to that in the TiO$_2$ scaffold. Additionally, the observation of the large J-V hysteresis in the PSCs based on CsPbI$_3$ that was not a ferroelectric material and did not contain organic polar molecule suggested that ferroelectricity was not the origin of the J-V hysteresis in PSCs.

As stated earlier, lowering the crystal size of CsPbI$_3$ could induce lattice strain to partially stabilize the α-CsPbI$_3$ phase. To further stabilize the α-CsPbI$_3$ phase, Swarnkar et al. have put a step forward to exploit nanocrystal surfaces to stabilize the α-CsPbI$_3$ phase and lead-based CsPbBr$_3$, and after the synthesis of CsPbI$_3$ nanocrystals (NCs), a new extraction solvent, methyl acetate (MeOAc), was applied for purification. This antisolvent removed excess unreacted precursors without inducing agglomeration, possibly due to the isolation of the quantum dots (QDs) without full removal of the surface species, which made the NC phase stable for months in ambient storage. For device fabrication, the NC solution was spin coated on substrate followed by dipping in MeOAc-based solution for ligand removal. The PSCs with the device structure of FTO/c-TiO$_2$/CsPbI$_3$/Spiro-MeOTAD/MoO$_x$/Al (Fig. 6.5a,b) exhibited a considerably high PCE of 10.77% with a V_{oc} of 1.23 V (Fig. 6.5c). Soon after, additive engineering was applied by Wang et al. to reduce grain size for stabilizing the α-CsPbI$_3$ phase. By adding a small amount (1.5 wt%) of sulfobetaine zwitterion in CsPbI$_3$ precursor solution, small CsPbI$_3$ grains with average size of 30 nm were obtained due to the electrostatic interaction of zwitterion with the ions and colloids (Fig. 6.5d,e), which helped to enlarge grain surface area to stabilize α-CsPbI$_3$ phase. By further incorporating 6% Cl ions, the PSCs of ITO/PTAA$_2$/CsPb(I$_{0.98}$Cl$_{0.02}$)$_3$/PCBM/C$_{60}$/BCP/metal achieved a PCE of 11.4%.

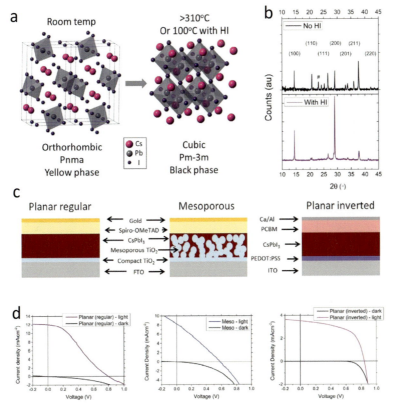

Figure 6.4 CsPbI$_3$ PSCs employing different cell structures. (a) Crystal structures of yellow and black phase CsPbI$_3$, (b) XRD patterns of the CsPbI$_3$ films deposited from the solution without and with the addition of HI, (c) scheme of different cell structures, and (d) J-V curves of the CsPbI$_3$ PSCs based on different cell structures. (Chen et al., 2018; Eperon et al., 2015).

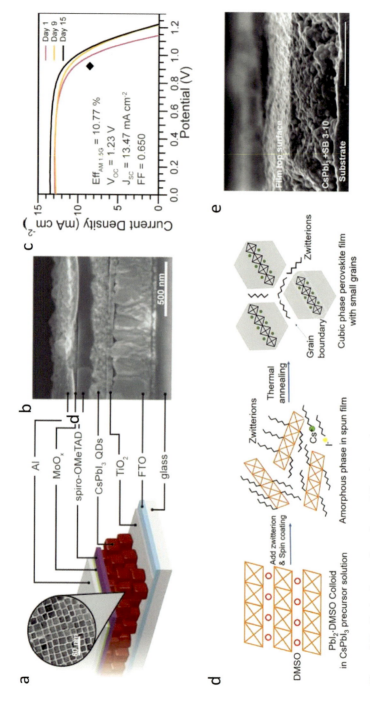

Figure 6.5 Grain size engineering stabilized α-CsPbI$_3$ phase. (a) Structure of the PSCs based on CsPbI$_3$ QDs layers and (b) J–V curves of the CsPbI$_3$ QDs PSCs after different storage periods. (d) Mechanism of α-CsPbI$_3$ stabilization by zwitterion, (e) cross-sectional SEM image of CsPbI$_3$ films with SB3-10 zwitterion. (Chen et al., 2018; Selig et al., 2017).

Lead-Based Inorganic Perovskites | 153

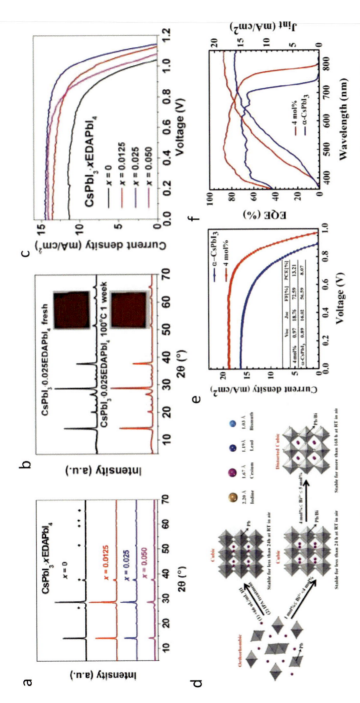

Figure 6.6 Doping stabilized α-CsPbI₃ phase. (a) XRD patterns of CsPbI₃ films containing different amount of EDAPbI₄, (b) XRD patterns of the CsPbI₃-0.025EDAPbI₄ films before and after 1 week storage at 100°C, and (c) J-V curves of the PSCs based on CsPbI₃ films containing different amounts of EDAPbI₄. (d) Schematic illustrating the change in crystal structures by HI/IPA treatment and Bi doping strategies, and (e) J-V curves and (f) IPCE spectra of the PSCs based on CsPbI₃ and Bi-doped CsPbI₃. (Chen et al., 2018).

In addition to grain size engineering, doping engineering has also been exploited to stabilize α-CsPbI$_3$. By incorporating a small amount of ethylenediamine (EDA) cation into CsPbI$_3$, Zhang et al. successfully avoided the undesirable formation of δ phase (Fig. 6.6a). The α-CsPbI$_3$ phase could be maintained at room temperature for months and at 100°C for >150 h (Fig. 6.6b). The PSCs of FTO/c-TiO$_2$/CsPbI$_3$·0.025EDAPbI$_4$/spiro-MeOTAD/Ag displayed a highly reproducible PCE of 11.8% (Fig. 6.6c). Besides organic ions, Hu et al. have doped the CsPbI$_3$ with Bi^{3+} ions, which helped to further stabilize the α-phase at room temperature. Several factors contributed to the enhancement in the phase stability of CsPbI$_3$ by Bi doping. First, the smaller radius of Bi^{3+} than that of Pb^{2+} afforded a larger tolerance factor for α-CsPb$_{1-x}$Bi$_x$I$_3$ (0.84) than that for α-CsPbI$_3$ (0.81). Second, Bi doping remarkably increased the microstrain in crystal structure and hence induced a slight distortion in the cubic structure. As previously reported, lattice strain was observed to induce crystal phase transition and act to shift the phase diagram for a material. Therefore, the presence of microstrain might contribute to the long-term phase stability of α-CsPb$_{1-x}$Bi$_x$I$_3$ (Fig. 6.6d). Besides, the incorporation of Bi^{3+} ions well reduced the grain size, which was also favorable for stabilizing α-CsPbI$_3$. Besides stability issues, Bi^{3+} doping narrowed the bandgap from 1.73 to 1.56 eV at 4 mol% and extended light absorption range to 795 nm (Fig. 6.6f). Fewer traps and defects were found in Bi-doped CsPbI$_3$. By employing TiO$_2$ compact layer (c-TiO$_2$) and CuI as ETM and HTM, respectively, the full inorganic PSCs obtained a PCE of 13.21% at the Bi^{3+} concentration of 4 mol% (Fig. 6.6e), and obviously higher stability has been demonstrated.

6.3.2 CsPbBr$_3$ Perovskite Solar Cells

By replacing the I with Br ions, CsPbBr$_3$ could be obtained, and similarly with MAPbX$_3$ series PVKs, the bandgap of CsPbBr$_3$ increased to 2.3 eV with respect to CsPbI$_3$ (1.73 eV). The orthorhombic phase CsPbBr$_3$ was stable at room temperature, so that the first demonstration of CsPbX$_3$ PVKs in PSCs was based on CsPbBr$_3$. In the work reported by Kulbak et al., the CsPbBr$_3$ layer was prepared by a two-step sequential method, in which the PbBr$_2$ layer was first deposited by spin coating, followed by chemical conversion in CsBr methanol solution. The PSCs with the device structure of FTO/m-TiO$_2$/CsPbBr$_3$/HTM/Au were fabricated and evaluated (Fig. 6.7a). The PSCs with the HTM of Spiro-MeOTAD, PTAA, and CBP achieved PCEs of 4.77–4.98, 5.72–5.95, and 4.09–4.72%, respectively, which were all comparable with the PSCs based on MAPbBr$_3$ (Fig. 6.7b) in their work. Furthermore, the PSCs without HTM exhibited a relatively high PCE of 5.32–5.47%, implying a high hole conductivity of CsPbBr$_3$, which was also similar with that of MAPbBr$_3$. Besides similar photovoltaic response, CsPbBr$_3$ exhibited an obviously higher stability than MAPbBr$_3$, especially for thermal stability (Fig. 6.7c), leading to the higher device stability for CsPbBr$_3$ device than that for MAPbBr$_3$ device (Fig. 6.7d). The analysis of the electron beam induced current (EBIC), which used an electron beam to act as a light source equivalent for generating electron–hole pairs in the junction area, indicated that CsPbBr$_3$ was stable but MAPbBr$_3$ was under degradation during multiple times scanning (Fig. 6.7e). Therefore, CsPbBr$_3$ PSCs were less prone to environmental degradation compared to MAPbBr$_3$ PSCs due to the relatively high temperature and single-phase degradation process of CsPbBr$_3$.

The application of colloidal semiconductor nanocrystals (NCs) in constructing optoelectronics is attractive, which enables solution processing and provides a powerful platform for tuning optical and electrical properties. However, it is a challenge to convert NC solutions to high-quality NC films preserving their properties, because the ligands and solvents with long alkyl chains would prevent the production of dense films and form an insulating layer around the NCs. To solve this problem, Akkerman et al. reported a novel and large-scale approach by using short, low-boiling-point ligands (propionic acid (PrAc), and butylamine (BuAm)) and environmentally friendly solvents (isopropanol (IPrOH) and hexane (HEX)) for the synthesis of CsPbBr$_3$ NCs (Fig. 6.8a,b). The final NC solution exhibited a 58 ± 6% photoluminescence quantum yield (PLQY), which was lower than those synthesized from long alkyl chains. However, the PLQY of the thin film deposited from these NCs (35 ± 4%) is higher than those from long alkyl chains (≈30%) that remained non-conductive. Besides, an amplified spontaneous emission (ASE) threshold as low as 1.5 µJ/cm^2 was detected, which was the lowest value reported for CsPbBr$_3$ NCs and overall for PVK thin films. Such ASE threshold was also at the lowest levels among

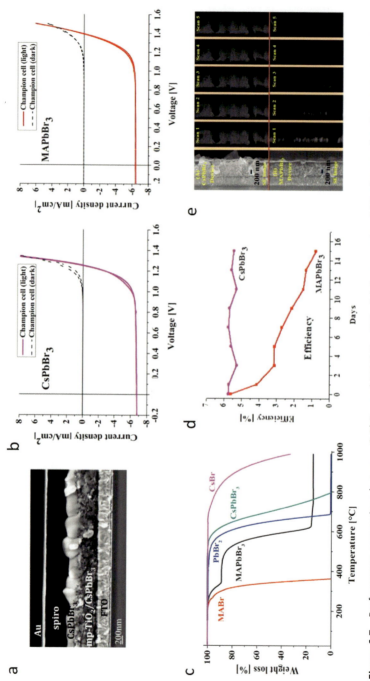

Figure 6.7 Performance comparison between CsPbBr₃ and MAPbBr₃ PSCs. (a) Cross-sectional SEM image of CsPbBr₃ PSCs. Reproduced with permission. Copyright 2015, American Chemical Society. (b) J-V curves of CsPbBr₃ and MAPbBr₃ PSCs, (c) thermogravimetric analyses of different PVK materials, (d) change in the PCE with time for CsPbBr₃ and MAPbBr₃ PSCs, (e) repetitive sequential EBIC responses of the cross sections of CsPbBr₃ and MAPbBr₃ PSCs. (Chen et al., 2018).

Lead-Based Inorganic Perovskites | 157

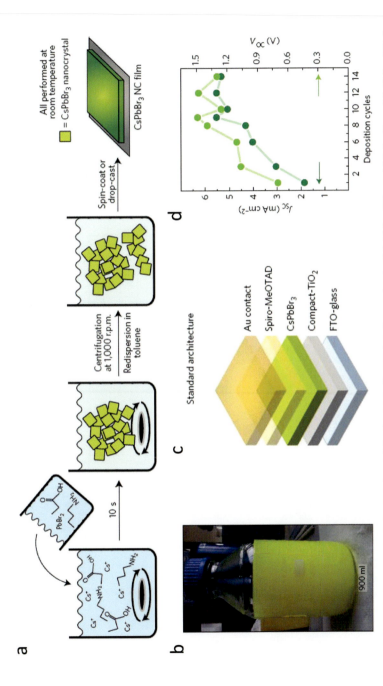

Figure 6.8 PSCs based on CsPbBr₃ QDs layer. (a) Schematic illustrating the synthesis of CsPbBr₃ QDs and fabrication of CsPbBr₃ QDs layer, (c) photo of scaled-up 2 g synthesis of CsPbBr₃ QDs, (d) cell structure of the PSCs based on CsPbBr₃ QDs layer, and (d) change in the J_{sc} (dark green dots) and V_{oc} (light green dots) of CsPbBr₃ PSCs with NC deposition cycle number. (Chen et al., 2018).

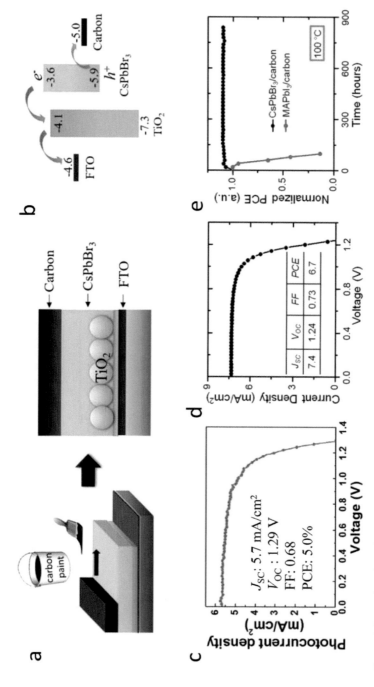

Figure 6.9 Carbon-based CsPbBr₃ PSCs. (a) Scheme of the carbon deposition process, (b) working principle in C-PSCs, and (c) J–V curve of CsPbBr₃ C-PSCs. (d) J–V curve of CsPbBr₃ C-PSCs and (e) PCE change with time for CsPbBr₃ and MAPbI₃ C-PSCs under 100°C. (Chen et al., 2018; Chang et al., 2016).

all kinds of inorganic NCs, including the inorganic NCs commonly used in laser devices. The NC film was applied in PSCs by employing a typical device structure of FTO/c-TiO$_2$/CsPbBr$_3$/Spiro-MeOTAD/Au (Fig. 6.8c). After optimizing the film thickness by adjusting spin-coating cycles, the PSCs achieved a PCE of 5.4% with a V_{oc} as high as 1.5 V (Fig. 6.8d). Such V_{oc} has been the highest value for Cs-based PSCs, which was expected to be further increased by designing the charge-extracting layers.

Though HTM-based CsPbBr$_3$ PSCs have demonstrated a relatively enhanced stability compared with their organic counterpart, the presence of organic HTM still limited the device stability. Since I-PVKs have similar optoelectronic properties with OIH-PVKs, I-PVKs could also act as both a light harvester and a hole transporter without using organic HTM. Both Au and carbon could be exploited as hole-extraction electrodes in HTM-free PSCs. However, compared with Au electrode, carbon is cheap, stable, inert to ion migration originating from perovskite and metal electrodes, inherently water resistant, which is more promising for HTM-free I-PSCs. Chang et al. have, for the first time, applied carbon electrode as hole-extraction electrode without using organic HTM and noble metal electrode (C-PSCs). As shown in Fig. 6.9a,b, carbon electrode was directly deposited on the CsPbBr$_3$ layer by painting a carbon paste, followed by annealing at low temperature. Through a systematical optimization, the CsPbBr$_3$ C-PSCs achieved a PCE of 5.0% with a V_{oc} of 1.29 V (Fig. 6.9c). Soon after, a similar work was reported by Liang et al., which got an enhanced PCE of 6.7% for small active area (Fig. 6.9d) and 5.0% for large active area (1.0 cm^2). Very similarly, both works indicated that CsPbBr$_3$ C-PSCs have obviously higher stability than that of MAPbI$_3$ C-PSCs, especially for thermal stability (Fig. 6.9e).

6.3.3 CsPbI$_{3-x}$Br$_x$ Perovskite Solar Cells

The successful demonstration of the PSCs with CsPbI$_3$ and CsPbBr$_3$ as absorbers has encouraged researchers to pay attention on preparing mixed halide CsPbI$_{3-x}$Br$_x$ for PSCs, especially when the black α-CsPbI$_3$ phase is not stable at room temperature and the bandgap of CsPbBr$_3$ is too large. Almost at the same time, Sutton et al. and Beal et al. investigated CsPbI$_{3-x}$Br$_x$ materials with the x ranging from 0 to 3 and found that the bandgap decreased as x value increased (Fig. 6.10a,b). Importantly, the partial substitution of I with Br ions contributed to a stable PVK phase at room temperature. Therefore, $CsPbI_{3-x}Br_x$ has shown much promise for tandem PV application as the stable $CsPbI_{3-x}Br_x$ with the bandgap of about 1.9 eV was obtained. Both works have compared the stability of $CsPbI_2Br$ with the organic counterparts, which demonstrated $CsPbI_2Br$ could stand for a longer time than $MAPbX_3$ under high temperature (85–180°C). The PSCs with the device structure of ITO/PEDOT:PSS/CsPbI$_2$Br/PCBM/BCP/Al achieved a PCE of 6.8% with a V_{oc} of 1.12 V and a J_{sc} of 10.9 mA/cm², while the PSCs with the structure of FTO/c-TiO$_2$/CsPbI$_2$Br/Spiro-OMeTAD/Ag exhibited a much higher performance (PCE = 9.8%, V_{oc} = 1.11 V, and J_{sc} = 11.89 mA/cm²). As expected, these promising results have been attracting more and more attention on CsPbI$_2$Br PSCs.

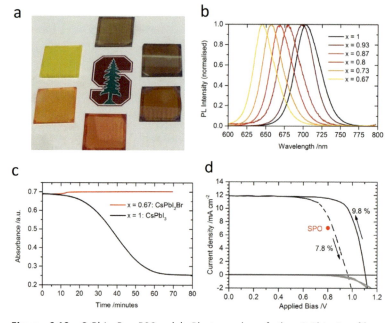

Figure 6.10 $CsPbI_{3-x}Br_x$ PSCs. (a) Photographs of the $CsPbI_{3-x}Br_x$ films. (b) PL spectra of the $CsPb(I_xBr_{1-x})_3$ films with different x values, (c) change in the absorbance of CsPbI$_3$ and CsPbI$_2$Br film with time, and (d) J-V curves of the PSCs based on CsPbI$_2$Br film. (Chen et al., 2018).

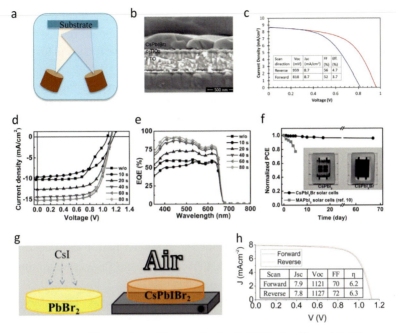

Figure 6.11 Different deposition methods of CsPbI$_{3-x}$Br$_x$ films. (a) Scheme of vapor evaporation processes, (b) SEM image of the CsPbIBr$_2$ film deposited by vapor process, and (c) J-V curves of the HTM-free PSCs based on CsPbIBr$_2$ film. Effect of the vapor time on the performance ((d) J-V curves and (e) IPCE spectra) of the CsPbI$_2$Br PSCs, and (f) change in the normalized PCE of MAPbI$_3$ and CsPbI$_2$Br PSCs with time. Insets in (f) were the photographs of the CsPbI$_3$ and CsPbI$_2$Br PSCs after storage for two weeks. (g) Schematic illustrating the conversion of PbBr$_2$ to CsPbIBr$_2$ and (h) J-V curves of the CsPbIBr$_2$ PSCs. (Chen et al., 2018).

The promising stability and decent performance have attracted much attention, and many strategies have been developed to prepare CsPbI$_{3-x}$Br$_x$ films for PSCs, especially when the solubility of Br ions in DMF or DMSO was too low to deposit the CsPbI$_{3-x}$Br$_x$ film with large film thickness. Thermal evaporation is an important method (Fig. 6.11a), which has been used to prepare relatively high-quality CsPbIBr$_2$. To our knowledge, thermal evaporation was first used to prepare CsPbIBr$_2$ film for PSCs by Ma et al. CsI and PbBr$_2$ with the same molar quantity were evaporated onto the substrates with different temperatures, followed by post-annealing at a different temperature. CsPbIBr$_2$ crystals with the size as large as 500–1000 nm were obtained when the substrate and annealing temperatures were set at 75°C and 250°C (Fig. 6.11b), respectively. HTM-free PSCs were prepared by directly depositing Au electrode on the CsPbIBr$_2$ film, which exhibited PCEs of 4.7% and 3.7% for reverse and forward scans, respectively (Fig. 6.11c).

In order to accurately control the stoichiometric ratios of the precursors in the CsPbI$_{3-x}$Br$_x$ films, the hygroscopic nature of CsI and CsBr was carefully considered in the work reported by Chen et al. The successful co-deposition of the stoichiometrically balanced PbI$_2$ and CsBr leaded to a high-quality CsPbI$_2$Br film. Through controlling the post-annealing durations, crystal size could be controlled, and 60 s was enough to increase the grain size to ≈3 μm. The PSCs with the device structure of ITO/ Ca/C$_{60}$/CsPbI$_2$Br/TAPC/TAPC:MoO$_3$/Ag achieved a PCE of 11.8% (Fig. 6.11d,e). Similar to previous works, the vacuum-deposited CsPbI$_2$Br PSCs showed better stability than the MAPbI$_3$ counterpart, with 96% of the peak PCE was maintained after more than 2-month storage (Fig. 6.11f).

As an industrial technique, spraying process has obvious advantages over traditional spin-coating method, especially for scaling-up purpose. As a result, Lau et al. have exploited the spraying process to deposit CsPbI$_{3-x}$Br$_x$ films for PSCs, in which the CsI layer was sprayed on a pre-deposited PbBr$_2$ film for post-conversion (Fig. 6.11g). This strategy could also avoid the possible dissolution of PbBr$_2$ and CsPbIBr$_2$ film in the CsI solution during the conversion process at the second step. A systematical investigation on the effects of substrate temperature during spraying and the post-annealing temperature on the PVK quality was conducted. The optimized device with the structure of FTO/ m-TiO$_2$/CsPbIBr$_2$/Spiro-MeOTAD/Au produced a PCE of 6.3% (Fig. 6.11h).

In addition to the deposition method, doping strategies have been exploited to modify the optoelectronic properties of CsPbI$_{3-x}$Br$_x$. By partially substituting Cs$^+$ ions with K$^+$ ions, Nam et al. successfully improved the electron lifetime in Cs$_{1-x}$K$_x$PbI$_2$Br without compromising light absorption range, which helped to obviously increase the PCE (device structure: FTO/c-TiO$_2$/Cs$_{1-x}$K$_x$PbI$_2$Br/Spiro-OMeTAD/Au) to around 10%. Furthermore, the device stability was also improved in ambient atmosphere. Lau et al. have incorporated a small amount of Sr^{3+} ions into CsPbI$_2$Br for forming CsPb$_{1-x}$Sr$_x$I$_2$Br PVKs. As indicated, the surface of the perovskite film was enriched with Sr, which played a role in passivating defects in the films. As a result, charge recombination was suppressed and electron lifetime was increased, which significantly boosted PCE from 7.7% to 11.2% for the device of FTO/mp-TiO$_2$/CsPb$_{1-x}$Sr$_x$I$_2$Br/P3HT/Au. Very recently, Liang et al. have partially substituted Pb^{2+} ions in CsPbIBr$_2$ with Sn^{2+} ions. CsPb$_{0.9}$Sn$_{0.1}$IBr$_2$ had a bandgap of 1.79 eV with high stability. The C-PSCs with the structure of FTO/m-TiO$_2$/CsPb$_{0.9}$Sn$_{0.1}$IBr$_2$/C obtained a PCE of 11.33% and exceptional stability.

The formation of halide alloy materials with the composition of CsPbBr$_x$I$_{3-x}$ has exhibited considerably high phase stability and achieved decent device performance. Just like the organic counterpart, the phenomenon of light-induced dealloying has also been observed in CsPbI$_{3-x}$Br$_x$. Niezgoda et al. have studied the impact of halide dealloying on the performance of CsPbI$_2$Br PSCs. Under continuous light illumination, the device of ITO/c-TiO$_2$/CsPbI$_2$Br/Spiro-MeOTAD/Ag exhibited an extreme device performance evolution with the FF and J_{sc} increasing gradually with illumination time (Fig. 6.12a). Accompanying with performance evolution, a reversible dealloying process was found from XRD and PL measurements (Fig. 6.12b), indicating a close relation with performance improvement and dealloying. Further study implied that holes became more mobile in dealloyed CsPbBrI$_2$, which facilitated hole injection into Spiro-MeOTAD. Therefore, the enhancement in FF and J_{sc} after dealloying should be due to the improved charge transport and collection. By combining PL, cathodoluminescence (CL), and transmission electron microscopy (TEM), Li et al. directly probed the phase segregation phenomenon in CsPbIBr$_2$. Under light and electron beam illumination, "iodide-rich" phases tended to form at grain boundaries and segregate as clusters inside the film (Fig. 6.12c,d). As suggested, phase segregation would induce the formation of a high density of mobile ions that moved along grain boundaries to pile up at the CsPbIBr$_2$/TiO$_2$ interface, which would form an injection barrier and electron accumulation for inducing large J-V hysteresis in planar PSCs. However, "iodide-rich" phases at grain boundary may also lead to a "bulk-heterojunction," which would help to accelerate charge separation and collection, and finally benefit the device performance.

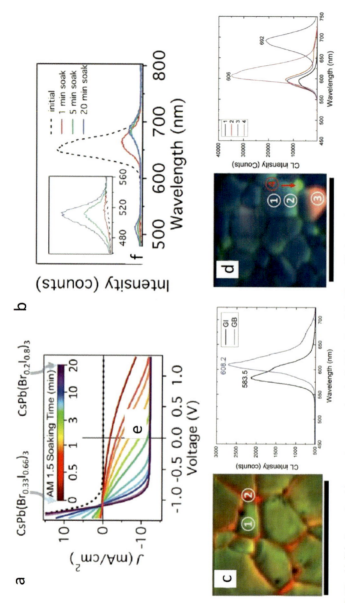

Figure 6.12 Halide ion migration and segregation in CsPbI$_{3-x}$Br$_x$. (a) Effect of light-soaking time on J-V curves for CsPbI$_{3-x}$Br$_x$ PSCs and (b) photoluminescence (PL) spectra of a CsPbI$_2$Br film on glass with different light illumination time. (c) CL spectrum mapping (acceleration voltage of 2 kV) with a superposition of different spectral windows of 530–590 and 590–640 nm and corresponding CL spectra for the areas 1 and 2. (d) CL spectrum mapping (acceleration voltage of 5 kV) with a superposition of different spectral windows of 530–630 and 630–730 nm and corresponding CL spectra for the areas 1, 2, 3, and 4. (Chen et al., 2018).

6.4 Lead-Free Inorganic Perovskites

6.4.1 CsSnI₃ Perovskite Solar Cells

Substitution of Pb with Sn ion in CsPbI$_3$ leads to CsSnI$_3$, which has two polymorphs at room temperature: B-γ-CsSnI$_3$ (a black orthorhombic phase, Fig. 6.13a) and Y-CsSnI$_3$ (a yellow phase with one-dimensional double-chain structure). The bandgap of B-γ-CsSnI$_3$ is near optimal for PV applications ($\approx$1.3 eV), with high optical absorption coefficient ($\approx 10^4$ cm^{-1}) and low exciton binding energy ($\approx$18 meV), which makes CsSnI$_3$ promising for PSCs. Because of its good p-type conductivity under Sn insufficient condition, CsSnI$_3$ was first used by Chung et al. as solid electrolyte in DSSCs with a ruthenium dye as the main light absorber, in which the CsSnI$_3$ was synthesized using a vacuum melt process at 450°C (Fig. 6.13b). Though CsSnI$_3$ was mainly considered a solid electrolyte for hole transport, the obviously enhanced spectral response at long wavelength range implied that CsSnI$_3$ should absorb and convert visible light for photocurrent enhancement (Fig. 6.13c). Almost at the same time, a Schottky photovoltaic device (ITO/CsSnI$_3$/Au/Ti) using CsSnI$_3$ as the light absorber was demonstrated by Chen et al., as illustrated in Fig. 6.13d,e. The CsSnI$_3$ layer in this device was deposited by the vacuum evaporation method, in which the alternating layers of CsI and SnCl$_2$ were sequentially deposited, followed by annealing at 175°C. This type of PSCs obtained a low PCE of 0.88% (Fig. 6.13f).

The promising bandgap and decent photovoltaic response inspired many researcher. Kumar et al. have demonstrated the first PSCs using the CsSnI$_3$ layer as light absorber (device structure: FTO/m-TiO$_2$/CsSnI$_3$/HTM/Au, Fig. 6.14a). The CsSnI$_3$ layer was deposited by one-step solution method at low temperature (70°C), and it was found that the solvent greatly affected the morphology of the CsSnI$_3$ layer. Compared with DMF and 2-methoxyethanol solvent, DMSO facilitated the formation of a more complete coverage and good pore filling on TiO$_2$ scaffold. Compared with Spiro-MeOTAD, 4,4′,4″-tris(N,N-phenyl-3-methylamino) triphenylamine (m-MTDATA) as HTM exhibited a superior overall performance due to higher oxidation potential. Since high V_{Sn} concentration is a dominant problem for generating high background carrier concentration and high charge recombination, SnF_2 has been exploited to address this issue, which helped to reduce V_{Sn} concentration without entering the $CsSnI_3$ lattice. An addition of 20% SnF_2 considerably boosted the photovoltaic performance, and a PCE of 2.02% was obtained with a J_{sc} of 22.70 mA/cm², a V_{oc} of 0.24 V, and an FF of 0.37 (Fig. 6.14b,c). An investigation of mechanism suggested that there was no significant energy barrier in the device, but a trap-assisted recombination mechanism should be the main cause for low photovoltaic performance.

As the film quality and V_{Sn} concentration greatly limited the performance of $CsSnI_3$ PSCs, Wang et al. have developed a simple solution-based method to address these issues. The solvent used for the precursor solution is a mixture consisting of methoxyactonitrile, DMF, and acetonitrile, which was used to dissolve the $CsSnI_3$ raw material. Film quality and V_{Sn} were controlled by varying post-treatment temperature (Fig. 6.14d–f). Several device structures of PSCs have been constructed for performance evaluation. The PSCs using mesoporous scaffolds (TiO_2 or Al_2O_3) exhibited considerably low PCEs (≈0% for TiO_2 and 0.32% for Al_2O_3), which might be due to constrained crystallization environment scaffolds, poor pore filling, and low-crystallinity impurities. By exploiting a planar device structure of ITO/c-TiO_2/$CsSnI_3$/Spiro-MeOTAD/Au, the performance was well improved with a PCE of 0.77%. The Spiro-MeOTAD used in this device was undoped, which would limit hole conductivity, because the necessary oxidation step for doped Spiro-MeOTAD would degrade the γ-$CsSnI_3$ phase. To solve this problem, the inverted planar PSCs using NiO_x as HTM was fabricated (Fig. 6.14g), which significantly promoted the performance. The investigation of the effects of annealing temperature on device performance indicated that the optimal temperature was 150°C (Fig. 6.14i). This was because though higher temperature could enlarge crystals for reducing charge recombination, the increase in V_{Sn} and film roughness would reduce carrier lifetime and deteriorate interface contact, respectively.

Due to the susceptibility of $CsSnI_3$ to oxidation, the low defect formation energy and a lot of pinholes, $CsSnI_3$ PSCs usually presented low performance and poor stability. To solve this problem, Marshall et al. conducted a systematical study by adding tin halide into

CsSnI$_3$ precursor solutions. First, they investigated the influence of the excess SnI$_2$ in the CsSnI$_3$ solution on CsSnI$_3$ film morphology and device performance. Compared with bare CsSnI$_3$ solution, the addition of the excess SnI$_2$ (10 mol%) well increased the performance of CsSnI$_3$ PSCs. This could be rationalized in terms of a reduction in background carrier density with excess SnI$_2$ that gave rise to recombination loses because it was known that the density of V_{Sn} defects (i.e., the primary source of the background carrier density) was suppressed when the PVK was synthesized in an Sn-rich environment. Besides, the positive vacuum level shift at the interface may also contribute to the increased V_{oc}. In combination with using CuI and ICBA as HTM and ETM, respectively, a PCE of 2.76% was obtained with a V_{oc} of 0.55 V.

The effects of SnBr$_2$, SnCl$_2$, and SnF$_2$ addition in CsSnI$_3$ have also been well evaluated. It was indicated that SnCl$_2$ has the most pronounced effects on CsSnI$_3$ film and PSCs. Since the morphology of the CsSnI$_3$ film prepared from SnCl$_2$-contained solution was not better than those prepared from SnI$_2$, SnBr$_2$, and SnF$_2$ solutions (Fig. 6.15a,b), morphology improvement could not account for the obvious performance enhancement. More detailed investigation demonstrated that the formation of a thin SnCl$_2$ layer on CsSnI$_3$ film surface was the predominant reason. The formation of SnCl$_2$ instead of CsSnCl$_3$ or mixed halide CsSnI$_{3-x}$Cl$_x$ may be due to the large difference in ionic radii between Cl and I (1.81 Å for Cl and 2.2 Å for I), and/or the very different structures of monoclinic CsSnCl$_3$ and orthorhombic CsSnI$_3$ at room temperature. During air exposure, the surface SnCl$_2$ thin layer sacrificed itself to form stable hydrate as well as SnO$_2$, which prevented underlying CsSnI$_3$ crystals from oxidation. As proved in this work, the excess SnCl$_2$ at the surface of the CsSnI$_3$ crystals tended to moderately n-dope the fullerene layer. As a result, a Schottky barrier would be formed to suppress the unwanted electron extraction from the fullerene to the ITO electrode at the pinhole sites. Consequently, charge recombination at the exposed ITO sites was reduced, boosting the PCE of CsSnI$_3$ to 3.56% (Fig. 6.15c,d). The positive effect of the SnCl$_2$ surface layer on device stability was well proved by the stability tests, as shown in Fig. 6.15e.

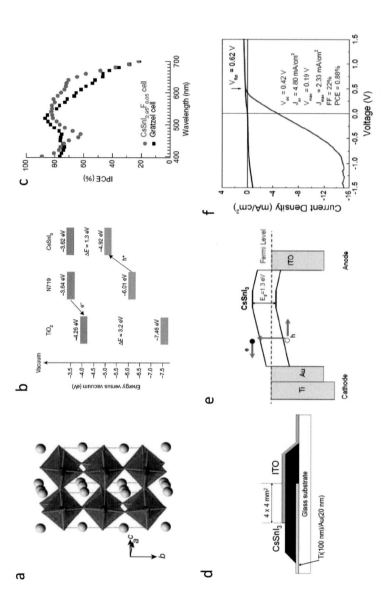

Figure 6.13 Primary CsSnI₃-based PV devices. (a) Crystal structure and (b) band structure CsSnI₃, and (c) IPCE spectra of DSSCs based on CsSnI₂.₉₅F₀.₀₅ and liquid electrolytes. Reproduced with permission. Copyright 2012, Nature Publishing Group. (d) Cell structure, (e) band alignment, and (f) J-V curves of CsSnI₃-based Schottky solar cells. (Chen et al., 2018; Chung et al., 2012).

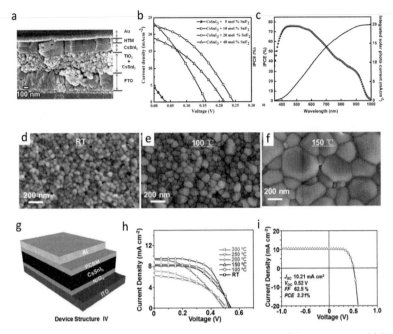

Figure 6.14 CsSnI₃ PSCs. (a) Cross-sectional SEM image, (b) J-V curves and (c) IPCE spectrum of CsSnI₃ PSCs. SEM images of CsSnI₃ films annealed at (d) RT, (e) 100°C, and (f) 150°C; (g) cell structure of CsSnI₃ PSCs; (h) J-V curves of the PSCs with the CsSnI₃ annealed at different temperature; and (i) J-V curve of champion CsSnI₃ PSCs. (Chen et al., 2018).

6.4.2 CsSnBr₃ Perovskite Solar Cells

Similar to CsPbX₃ (X = I, Br, and Cl) PVKs, the bandgap of CsSnX₃ also depended on the X elements. By substituting the I with Br, the bandgap of CsSnX₃ PVKs could be enlarged from 1.23 to 1.75 eV (Fig. 6.16a), which shows the promise to increase the V_{oc} of PSCs. Gupta et al. prepared CsSnBr₃ by one-step solution method to be used as an active absorber for the PSCs with "n-i-p" structure, and different ETM and HTM have been chosen for photovoltaic investigation. It was found that the most suitable ETM and HTM were m-TiO₂ and Spiro-MeOTAD, respectively. The effects of SnF₂ addition on film properties and device performance were thoroughly studied, which could considerably boost the photovoltaic performance with PCE increasing from only 0.01% to over 2% (Fig. 6.16c). Besides

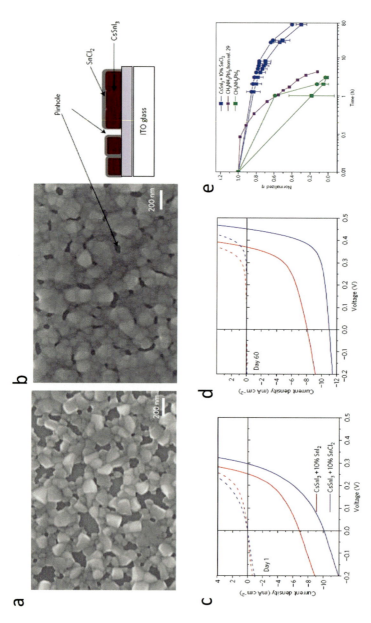

Figure 6.15 HTM-free CsSnI$_3$ PSCs. SEM images of CsSnI$_3$ films with (a) no tin halide additive and (b) 10 mol% added SnCl$_2$. J-V curves of the CsSnI$_3$ PSCs (c) before and (d) after a period of extended storage under nitrogen. (e) Stability tests of the non-encapsulated CsSnI$_3$ and MAPbI$_3$ PSCs with the same architecture under 1 sun constant illumination in ambient air. (Chen et al., 2018).

Lead-Free Inorganic Perovskites | 171

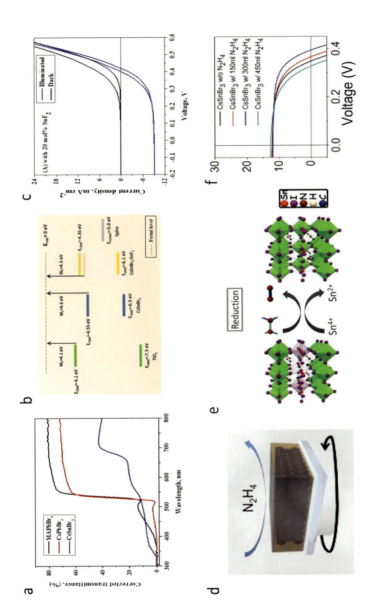

Figure 6.16 CsSnBr$_3$ PSCs. (a) UV-VIS absorption spectra of CsSnBr$_3$, CsPbBr$_3$, and MAPbBr$_3$, (b) band structure of CsSnBr$_3$ and CsSnBr$_3$:SnF$_2$, and (c) J-V curves of the PSCs based on CsSnBr$_3$:SnF$_2$ (20 mol%). (d) Reducing vapor atmosphere procedure for preparing CsSnBr$_3$, (e) proposed possible mechanism of hydrazine vapor reaction with Sn-based perovskite materials, and (f) J-V curves of CsSnBr$_3$ PSCs without and with various hydrazine vapor concentrations. (Chen et al., 2018).

the well-known reduction in background carrier density due to the reduction in V_{Sn}, the addition of SnF_2 was found to decrease the work function and make CB and VB more close to the CB of TiO_2 and VB of Spiro-MeOTAD, respectively, which facilitated the charge transfer at interfaces (Fig. 6.16b). Furthermore, the addition of SnF_2 well prevented Sn oxidation from X-ray irradiation and increased the stability in inert atmosphere.

For $CsSnX_3$ PVKs, Sn^{2+} ions have a large tendency to be oxidized to Sn^{4+} ions, which would form leakage pathways from p-type defect states, deteriorate device reproducibility, and limit PCE. In order to solve this problem, Song et al. have designed an effective process, in which a reducing vapor atmosphere (N_2H_4) was exploited during the deposition of $CsSnX_3$ by spin coating, as illustrated in Fig. 6.16d. The possible reaction path during the film deposition was suggested as $2SnI_6^{2-} + N_2H_4 \rightarrow 2SnI_4^{2-} + N_2 + 4HI$ (Fig. 6.16e), which resulted in more than 20% reduction in Sn^{4+}/Sn^{2+} ratios and, hence, well-suppressed carrier recombination. As a result, the PCE of $CsSnI_3$ PSCs was dramatically improved from ≈0.16 to 1.50%, while the PCE improvement of $CsSnBr_3$ PSCs was boosted from ≈2.36 to 2.82% with the highest PCE of 3.04%.

6.4.3 CsSnI$_{3-x}$Br$_x$ Perovskite Solar Cells

$CsSnI_3$ has been demonstrated as a lead-free halide PVK for functioning as a light absorber with high photocurrent densities. However, the PCEs were bottlenecked by low V_{oc}. To overcome this, Sabba et al. have modulated the V_{oc} by chemically doping $CsSnI_3$ with Br for forming CsSnI$_{3-x}$Br$_x$ ($0 \leq x \leq 3$). As Br doping concentration increased, the onset of optical bandgap edge transitions from 1.27 eV of $CsSnI_3$ to 1.37, 1.65, and 1.75 eV for $CsSnI_2Br$, $CsSnIBr_2$, and $CsSnBr_3$ (Fig. 6.17a,b), respectively, which was similar to the phenomenon of doping $MAPbI_3$ with Br. The Urbach energies (Uo) were deduced to be 16.8, 32, 39, and 32.6 meV for $CsSnI_3$, $CsSnI_2Br$, $CsSnIBr_2$, and $CsSnBr_3$, respectively, which suggested a low structural disorder in these materials. Obvious improvement in V_{oc} has been detected after Br incorporation, which was attributed to the decrease in V_{Sn}, as reflected by the lower charge carrier densities of 10^{15} cm^{-1} and a high resistance to charge recombination in CsSnI$_{3-x}$Br$_x$. By further adding SnF_2 (20 mol%) into CsSnI$_{3-x}$Br$_x$, V_{Sn} was well suppressed,

Lead-Free Inorganic Perovskites | 173

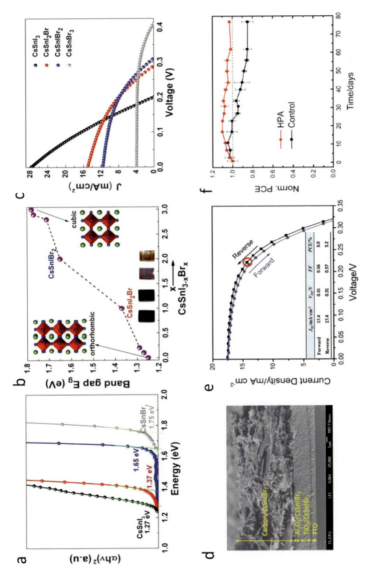

Figure 6.17 CsSnI$_{3-x}$Br$_x$ PSCs. (a) Tauc plots of the CsSnI$_{3-x}$Br$_x$ with different x values, (b) relation of the bandgap of CsSnI$_{3-x}$Br$_x$ with x values, and (c) J-V curves of CsSnI$_{3-x}$Br$_x$ PSCs. (d) Cross-sectional SEM image, (e) J-V curves of the meso C-PSCs based on CsSnIBr$_2$, and (f) change in efficiency with time for bare and HPA-modified CsSnIBr$_2$ C-PSCs. (Chen et al., 2018).

leading to a decrease in carrier densities and reduction in charge recombination. As a result, device performance was significantly enhanced especially for current density, and the CsSnI$_{2.9}$Br$_{0.1}$ PSCs yielded the highest PCE of 1.76% (Fig. 6.17c).

In order to obtain a stable PVK phase and reduce the bulk recombination induced by V_{Sn}, Li et al. have introduced hypophosphorous acid (HPA) into the CsSnIBr$_2$ precursor solution. The HPA additive not only acted as a complexant to promote the nucleation process, but also considerably reduced the carrier mobility and charge carrier density in CsSnIBr$_2$ films. When applied in PSCs, a C-PSC (HTM-free PSC) was constructed with the CsSnIBr$_2$ precursor solution being dropped into a multilayer mesoporous layer (TiO$_2$/Al$_2$O$_3$/C). The penetration of precursor solution in the mesoporous layers formed the final C-PSCs (Fig. 6.17d). Compared with the CsSnIBr$_2$ film from bare precursor solution, the HPA-induced CsSnIBr$_2$ film significantly improved the V_{oc} and FF of C-PSCs, boosting PCE to 3.2% (Fig. 6.17e). Furthermore, the synergistic effect of HPA addition and carbon electrode afford the CsSnIBr$_2$ C-PSCs with considerably high stability. That is, almost no PCE was decayed after 77 d (Fig. 6.17f) and the device maintained 98% of the initial PCE after 9 h continuous power output at 473 K.

6.4.4 CsGeI$_3$ Perovskite Solar Cells

CsGeX$_3$ has also been thought to be a potential light absorber in PSCs. Theoretical calculation implies that germanium halide PVKs have high absorption coefficients and similar absorption spectra and carrier transport properties as the lead analogues. The bandgap of CsGeX$_3$ depends on halide ion, and the bandgaps of CsGeCl$_3$, CsGeBr$_3$, and CsGeI$_3$ were calculated to be about 3.67, 2.32, and 1.53 eV, respectively (Fig. 6.18a,b). The bandgap of 1.53 eV made CsGeI$_3$ attractive as a promising absorber. However, due to the easy oxidation of Ge^{2+} cations to Ge^{4+} cations, the successful PSCs based on CsGeI$_3$ have been rarely reported. Krishnamoorthy et al. have, for the first time, reported PSCs based on CsGeI$_3$, in which a regular device structure was exploited with m-TiO$_2$ and Spiro-MeOTAD as ETM and HTM, respectively (Fig. 6.18c). A CsGeI$_3$ film with smooth morphology was prepared by applying DMF as the solvent of precursor solution. Such CsGeI$_3$ PSCs exhibited a J_{sc} of 5.7 mA/cm^2, as shown in Fig. 6.18d, which was encouragingly higher than that derived from the pristine Sn-based PVKs. However, the V_{oc} of this device was very low, which is possibly attributed to the oxidation of Ge^{2+} to Ge^{4+} during film synthesis and device fabrication. In addition, the ease of oxidation of CsGeI$_3$ also significantly limited the device stability, which even made photovoltaic test in air atmosphere impossible.

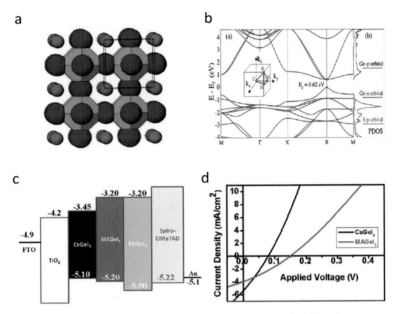

Figure 6.18 CsGeI$_3$ PSCs. (a) Crystal structure, (b) calculated band structure and projected density of states of cubic CsGeI$_3$, (c) schematic energy level diagram of CsGeI$_3$, MAGeI$_3$, and FAGeI$_3$, and (d) J-V curves of CsGeI$_3$ and MAGeI$_3$ PSCs. (Chen et al., 2018).

6.5 Perovskite-Derived Materials

6.5.1 Sn-Based Perovskites

Sn^{2+}-based PVKs are commonly faced with the ease of oxidation of Sn^{2+} to Sn^{4+} ions in ambient atmosphere. As a consequence, the formation of Sn^{4+} impedes the charge neutrality of the PVKs and causes the degradation of PVKs. To avoid oxidation, this type of PVKs could only be made in inert atmosphere and need rigorous encapsulation. Promisingly, Sn-based molecular iodosalt compounds

(A_2SnI_6) with the Sn ion at +4 oxidation state made them stable in air and moisture. Due to the near optimal bandgap of ≈1.3 eV and high absorption coefficient (over 10^5 cm^{-1} at 1.7 eV), Cs_2SnI_6, one type of molecular iodosalt compounds (Fig. 6.19a), has been well exploited for PV application. Lee et al. have, for the first time, applied this material in DSSCs as the solid-state electrolyte for hole transport (Fig. 6.19b,c). In combination with the efficient mixture of porphyrin dyes, a PCE of 8% has been achieved.

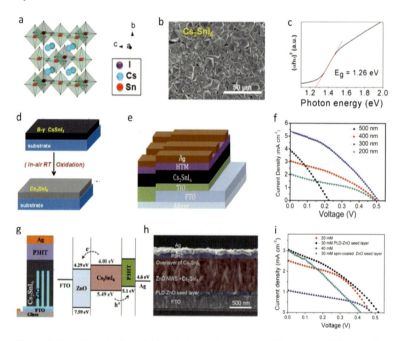

Figure 6.19 Cs_2SnI_6 PSCs. (a) Crystal structure, (b) SEM image, and (c) Tauc plot of Cs_2SnI_6. (d) Scheme of the conversion of B-γ-$CsSnI_3$ to Cs_2SnI_6 by oxidation, (e) cell structure of the PSCs based on Cs_2SnI_6, and (f) J-V curves of PSCs based on Cs_2SnI_6 with different thicknesses. (g) Scheme of cell structure and working principle, (h) cross-sectional SEM image, and (i) J-V curves of the PSCs based on Cs_2SnI_6 and ZnO NRs. (Chen et al., 2018).

Due to the promising semiconducting properties, the application of Cs_2SnI_6 in PSCs has been well exploited. Qiu et al. found that the spontaneous oxidative conversion of unstable B-γ-$CsSnI_3$ would lead to the formation of air-stable Cs_2SnI_6. They came out with an idea to exploit B-γ-$CsSnI_3$ as the precursor to obtain Cs_2SnI_6. As a result, they developed a thermal evaporation method to grow high-quality

B-γ-CsSnI$_3$, in which CsI and SnI$_2$ layers were successively deposited by thermal evaporation, followed by annealing in N$_2$ atmosphere, and the resulting B-γ-CsSnI$_3$ was automatically converted to Cs$_2$SnI$_6$ in air atmosphere, as illustrated in Fig. 6.19d. Then, the Cs$_2$SnI$_6$ film was, for the first time, applied in PSCs as a light absorber with the device structure of FTO/c-TiO$_2$/Cs$_2$SnI$_6$/P3HT/Ag (Fig. 6.19e), which achieved a PCE of 0.96% with a V_{oc} of 0.51 V and a J_{sc} of 5.41 mA/cm^2 after the systematical optimization of Cs$_2$SnI$_6$ thickness (Fig. 6.19f). Furthermore, the pronounced stability of Cs$_2$SnI$_6$ in ambient atmosphere led to considerably stable PSCs over 1 week. Soon afterward, they applied a one-step solution method that has been reported to deposit Cs$_2$SnI$_6$, in which Cs$_2$SnI$_6$ powder was first synthesized by solution process followed by dissolving in DMF as Cs$_2$SnI$_6$ precursor solution. By adopting ZnO nanorods (NRs) as ETM, the PSCs with the device structure of FTO/ZnO NRs/Cs$_2$SnI$_6$/P3HT/Ag could work well (Fig. 6.19g,h). After optimizing ZnO seed layers and NR morphology, a PCE of 0.86% was obtained with a V_{oc} of 0.52 V and a J_{sc} of 3.20 mA/cm^2 (Fig. 6.19i).

By substituting I with Br ions in Cs$_2$SnI$_6$, Lee et al. have changed the bandgap of Cs$_2$SnI$_{6-x}$Br$_x$ from 1.3 to 2.9 eV with increasing x value (Fig. 6.20a,b). This was realized by a two-step solution method with the deposition of CsI or CsBr film, followed by the chemical reaction with SnI$_4$ or SnBr$_4$ solutions. With increasing Br composition, the color of the Cs$_2$SnI$_{6-x}$Br$_x$ films changed from dark brown to light yellow due to the enlarged bandgap (Fig. 6.20c). This range of change in bandgap was obviously larger than that of CsSnX$_3$, which ranged from 1.3 eV (CsSnI$_3$) to 1.7 eV (CsSnBr$_3$). For solar cell assembly, a sandwich-type photochemical cell (FTO/bl-TiO$_2$/Sn-TiO$_2$/Cs$_2$SnI$_{6-x}$Br$_x$/Cs$_2$SnI$_6$(HTM)/LPAH/FTO) with an all-solid-state ionic conductor blended with inorganic materials (Cs$_2$SnI$_6$) and succinonitrile was constructed. The result of photovoltaic measurement indicated that the increase in Br content led to reduction in J_{sc}, with V_{oc} increasing gradually (Fig. 6.20d). At $x = 2$ (Cs$_2$SnI$_4$Br$_2$), the PSCs achieved the best device performance with a PCE of 2.02%, a V_{oc} of 0.563 V, and a J_{sc} of 6.225 mA/cm^2 and an FF of 0.58. The test on device stability demonstrated that Cs$_2$SnI$_4$Br$_2$ PSCs were more stable than Cs$_2$SnI$_6$ PSCs, which maintained almost their initial PCE after 50 days storage in air (Fig. 6.20e).

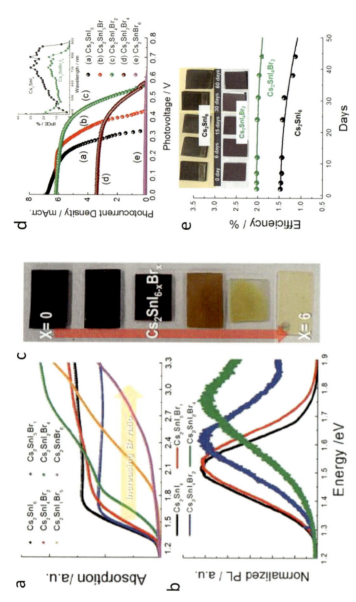

Figure 6.20 $Cs_2SnI_{6-x}Br_x$ PSCs. (a) UV-VIS absorption spectra, (b) PL spectra, and (c) photographs of $Cs_2SnI_{6-x}Br_x$ with different x values. (d) J–V curves of the PSCs based on $Cs_2SnI_{6-x}Br_x$ with different x values. Inset was the IPCE spectra of the PSCs based on Cs_2SnI_6 and $Cs_2SnI_4Br_2$. (e) Change in PCEs with time for the PSCs based on Cs_2SnI_6 and $Cs_2SnI_4Br_2$. Inset is the photograph of Cs_2SnI_6 and $Cs_2SnI_4Br_2$ films after storage for different time. (Chen et al., 2018).

6.5.2 Bi-Based Perovskites

Bi is an interesting replacement candidate for Pb and Sn. However, due to the different valence state, it is impossible for trivalent Bi^{3+} ion to directly replace Pb^{2+} or Sn^{2+} in the PVK structure. As a result, Bi-based PVKs exhibit a structure difference from traditional PVK structure and have a huge structural diversity with the dimension ranging from 0D dimer units, to 1D chain-like motifs, to 2D layered networks, and 3D double PVK. $Cs_3Bi_2I_9$ belongs to 0D dimer species, which consists of bioctahedral $(Bi_2I_9)^{3-}$ clusters surrounded by Cs^+ cations (Fig. 6.21a). Park et al. have exploited $Cs_3Bi_2I_9$ as a light absorber in PSCs, which presented a similar band structure with its counterparts of $MA_3Bi_2I_9$ with the bandgap of ≈2.2 eV (Fig. 6.21b). The inorganic nature and stable Bi^{3+} ions favor the deposition of $Cs_3Bi_2I_9$ materials in ambient atmosphere, and the $Cs_3Bi_2I_9$ film deposited by one-step method shows hexagonal thin sheet structure with the crystalline growth along the c-axis. The PSCs with the device structure of $FTO/m-TiO_2/Cs_3Bi_2I_9/Spiro-MeOTAD/Ag$ (Fig. 6.21c) obtained a PCE of 1.09% with a V_{oc} of 0.85 V and a J_{sc} of 2.15 mA/cm^2 (Fig. 6.21d). Obviously, high stability has been observed for this type of PSCs, and almost no hysteresis has been observed for the as-fabricated PSCs regardless of scanning rates. However, after 1-month storage, a large J-V hysteresis was observed, implying a possible interaction between $Cs_3Bi_2I_9$ with the additives in HTM. Following this work, Johansson et al. have synthesized $CsBi_3I_{10}$ and exploited it as a light absorber in PSCs. Differently, $CsBi_3I_{10}$ was shown to have a layered structure with a different dominating crystal growth direction than $Cs_3Bi_2I_9$. The absorption range extended to about 700 nm with a high coefficient of 1.4×10^5 cm^{-1} between 350 and 500 nm (Fig. 6.21e). The device of $FTO/m-TiO_2/CsBi_3I_{10}/P3HT/Ag$ obtained a PCE of 0.4% with a V_{oc} of 0.31 V and a J_{sc} of 3.4 mA/cm^2 (Fig. 6.21f), and the IPCE spectra covered the visible spectrum up to around 700 nm. However, time-dependent light absorption spectra indicated that the stability of the $CsBi_3I_{10}$ film was lower compared with $MAPbI_3$, which was a drawback of $CsBi_3I_{10}$.

180 | *Application in Perovskite Solar Cells*

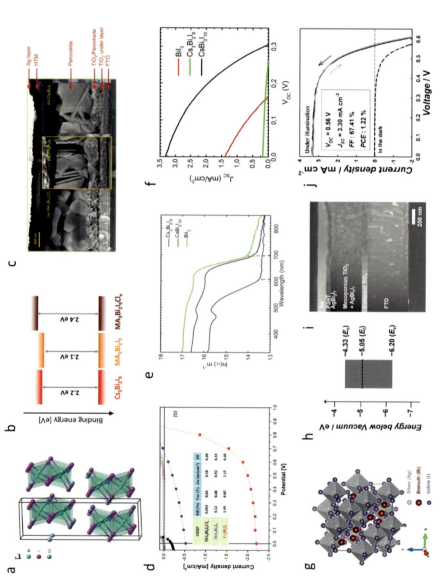

Figure 6.21 Bi-based PVKs for PSCs. (a) Crystal structure of $Cs_3Bi_2I_9$, (b) band structure of different Bi-based PVKs, (c) cross-sectional SEM image, and (d) J-V curves of $Cs_3Bi_2I_9$, $MA_3Bi_2I_9$, and $MA_3Bi_2I_9Cl_x$ PSCs. (e) UV-VIS absorption spectra of $Cs_3Bi_2I_9$, $CsBi_3I_{10}$, and BiI_3 films, and (f) J-V curves of $Cs_3Bi_2I_9$, $CsBi_3I_{10}$, and BiI_3 PSCs. (g) Crystal structure and (h) band structure of $AgBi_2I_7$, (i) cross-sectional SEM image, and (j) J-V curves of $AgBi_2I_7$ PSCs. (Chen et al., 2018; Kim et al., 2016).

Compared with low dimensional materials, 3D materials are more favorable for PV application in view of their semiconducting properties. Therefore, some researchers have focused on increasing the dimensionality of Bi-based PVK. Ag^+ and Cu^+ ions have been introduced into iodobismuthate materials based on bismuth iodide to obtain 3D structure. The silver–bismuth–iodine ternary system belongs to the silver iodobismuthate family, which could crystallize to $AgBi_2I_7$ (Fig. 6.21g). Kim et al. have reported a solution-based method to deposit air-stable $AgBi_2I_7$ using n-butylamine as the solvent to dissolve BiI_3 and AgI, which was determined to be a cubic structure with an E_g of 1.87 eV (Fig. 6.21h). The device of FTO/m-TiO$_2$/AgBi$_2$I$_7$/P3HT/Au was constructed (Fig. 6.21i), which obtained a J_{sc} of 3.30 mA/cm^2, a V_{oc} of 0.56 V, an FF of 0.67, and a PCE of 1.22% for the best device (Fig. 6.21j). The stability measurement indicated that the PCE of the best device remained above 1.13% for over 10 days of alternating device storage and testing under ambient conditions, suggesting the high air stability of $AgBi_2I_7$.

6.5.3 Sb-Based Perovskites

Similar to Bi-based PVKs, Sb-based I-PVKs have the basic structure of $A_3Sb_2X_9$ (A: Rb^+ and Cs^+; X: Cl, Br, and I) with a 0D dimer crystal structure (with fused bioctahedron) or a 2D layered crystal structure. The layered structure was calculated to be superior compared to dimer form due to the direct bandgap nature, higher electron and hole mobility, and better tolerance to defects because of higher dielectric constants. $Cs_3Sb_2I_9$ could be produced with the two types of structures according to preparation methods. The electronic band structure calculation indicated that the dimer and layered structures exhibited an indirect and direct bandgaps (2.4 versus 2.06 eV), respectively. Because the solution method usually leads to the generation of dimer structure, Saparov et al. have developed a two-step thermal evaporation approach to grow layered $Cs_3Sb_2I_9$ with a bandgap of 2.05 eV. By applying this $Cs_3Sb_2I_9$ film in PSCs, the device of FTO/c-TiO$_2$/Cs$_3$Sb$_2$I$_9$/PTAA/Au only exhibited significantly low performance with a V_{oc} of 0.3 V and a J_{sc} of 0.1 mA/cm^2, which should be attributed to the presence of a large amount of deep level defects in the layered $Cs_3Sb_2I_9$. Substituting Cs with Rb ions facilitated the synthesis of layered $Rb_3Sb_2I_9$ by a low-temperature solution-based method due to the size effect of the smaller A site ions (188 pm and 172 pm for Cs$^+$ and Rb$^+$, respectively) (Fig. 6.21a). Replacing Cs with Rb narrowed the bandgap for Rb$_3$Sb$_2$I$_9$, about 1.98 eV for direct transition (Fig. 6.22b), with an absorption coefficient of 1×10^5 cm^{-1}. Harikesh et al. have developed a one-step solution method to prepare layered Rb$_3$Sb$_2$I$_9$ film and applied it in the PSCs of FTO/m-TiO$_2$/Rb$_3$Sb$_2$I$_9$/poly-TPD/Au (Fig. 6.22c). Such PSCs achieved a PCE of 0.66% with a V_{oc} of 0.55 V and a J_{sc} of 2.12 mA/cm^2 (Fig. 6.22d), a significantly higher value than that based on Cs$_3$Sb$_2$I$_9$.

6.5.4 Double Perovskites

By introducing trivalent and monovalent cations into the B-sites of halide PVKs, halide double PVKs with the structure of A$_2$B^{1+}B^{3+}X$_6$ are formed. When the A site was occupied with an inorganic cation, an inorganic double PVK could be obtained. This class of I-PVKs consists of two types of octahedra alternating in a rock-salt face-centered cubic structure. So far, many inorganic double PVKs that have suitable bandgaps for photovoltaic application have been theoretically predicted. Some of these compounds demonstrated indirect bandgaps, while the others have direct bandgaps. Indirect double PVKs include Cs$_2$AgBiCl$_6$ (2.2–2.8 eV), Cs$_2$AgBiBr$_6$ (1.8–2.2 eV), Cs$_2$AgBiI$_6$ (1.6 eV), Cs$_2$CuBiX$_6$ (X = Cl, Br, I; E_g = 1.3–2.0 eV), Cs$_2$CuSbX$_6$ (X = Cl, Br, I; E_g = 0.9–2.1 eV), Cs$_2$AgBiX$_6$ (X = Cl, Br, I; E_g = 1.1–2.6 eV), Cs$_2$AuSbX$_6$ (X = Cl, Br, I; E_g = 0.0–1.3 eV), etc. Direct double PVKs include Cs$_2$InSbCl$_6$ (1.02 eV), Cs$_2$InBiCl$_6$ (0.91 eV), Rb$_2$CuInCl$_6$ (1.36 eV), Rb$_2$AgInBr$_6$ (1.46 eV), Cs$_2$AgInBr$_6$ (1.50 eV), etc.

Following the calculation, some researchers have exploited methods to synthesize inorganic double PVKs for characterizations. Slavney et al. have synthesized Cs$_2$AgBiBr$_6$ single crystals by solution route, which exhibited an indirect bandgap of 1.95 eV and a long room-temperature PL lifetime of 660 ns. Such Cs$_2$AgBiBr$_6$ had high defect tolerance and stability. Volonakis et al. have exploited a solid-state reaction strategy to grow Cs$_2$AgBiCl$_6$ powder whose indirect bandgap was characterized to be 2.2 eV. McClure et al. have synthesized Cs$_2$AgBiBr$_6$ and Cs$_2$AgBiCl$_6$ by both solid-state and solution methods, while both Cs$_2$BiAgCl$_6$ and Cs$_2$BiAgBr$_6$ single crystals have been synthesized by Filip et al. using a solution method. Recently, Du et al. have reported the synthesis of Cs$_2$BiAgBr$_6$ with the doping of In or Sb, which helped to modulate the bandgap and light absorption range.

Perovskite-Derived Materials | **183**

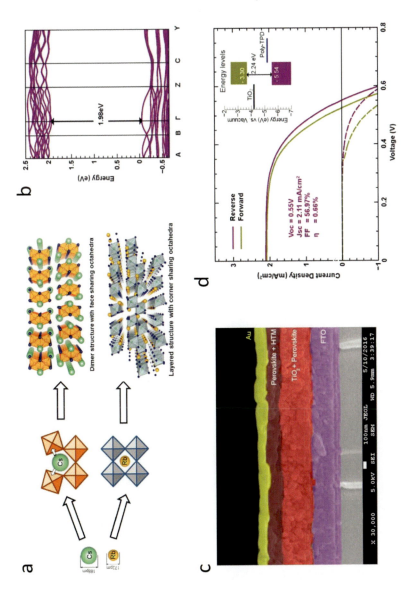

Figure 6.22 Sb-based PVKs for PSCs. (a) Crystal structure comparison between Cs$_3$Sb$_2$I$_9$ and Rb$_3$Sb$_2$I$_9$, (b) band structure of Rb$_3$Sb$_2$I$_9$, (c) cross-sectional SEM image, and (d) J-V curves of Rb$_3$Sb$_2$I$_9$ PSCs. Inset in (d) was the band structure of different materials in Rb$_3$Sb$_2$I$_9$ PSCs. (Chen et al., 2018; Harikesh et al., 2016).

Though many compounds have been theoretically predicted, only several compounds have been synthesized and characterized, and almost no suitable method has been developed to deposit thin film for photovoltaic application. Until recently, Greul et al. modified the spin-coating method to deposit $Cs_2AgBiBr_6$ films on TiO_2 mesoporous films, in which the preheated precursor solution was dropped onto the preheated and spinning substrate. Though the surface of $Cs_2AgBiBr_6$ films was rough with many aggregates, the substrate was fully covered, which favored the fabrication of PSCs. After depositing Spiro-MeOTAD as HTM, the best PCE of 2.24% with a V_{oc} of 1.06 V was achieved (Fig. 6.23c,d), which was also the first reported PCE for inorganic double PVKs.

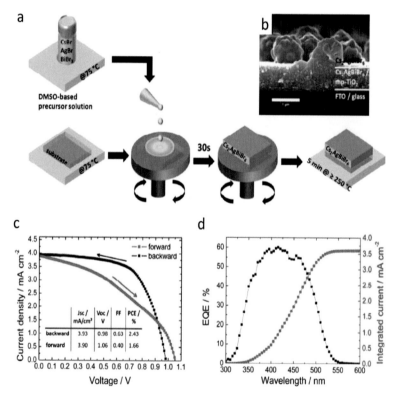

Figure 6.23 $Cs_2AgBiBr_6$ PSCs. (a) Scheme of the synthesis route for $Cs_2AgBiBr_6$ films, (b) cross-sectional SEM image of the $Cs_2AgBiBr_6$ films deposited on m-TiO_2, (c) J-V curve of the best performing device, and (d) EQE spectrum and integrated predicted current density of a PSC. (Chen et al., 2018; Greul et al., 2017).

6.6 Issues and Outlooks

In summary, I-PSCs have exhibited much promise as an alternative to OIH-PSCs and much progress has been achieved along this direction. However, the PCEs of I-PSCs are still significantly lower than those of OIH-PSCs. Some important issues need to be addressed to improve PCE, which represent the future directions.

1. Though I-PVK commonly exhibited high thermal stability, most of the efficient I-PVKs still suffer low stability in ambient atmosphere. The black α-CsPbI$_3$ phase spontaneously changes to the yellow CsPbI$_3$ phase at room temperature, while the Sn^{2+} ions in CsSnI$_3$ would be easily oxidized to Sn^{4+} ions in air atmosphere. To further improve the stability and performance of I-PSCs, these issues should be carefully resolved. Doping α-CsPbI$_3$ with Bi^{3+} ions has recently highlighted a promising way to stabilize the α-CsPbI$_3$ phase at room temperature, while to avoid oxidation of Sn^{2+} ions in CsSnI$_3$, device encapsulation should be carefully designed.
2. Compared with OIH-PVK, very limited synthesis strategies have been exploited to prepare I-PVK, especially for solution-based strategies, which have largely suppressed the enhancement in PCE. More efforts are in urgent demand to develop effective deposition methods to control the crystal growth and morphology of I-PVK films to improve the device performance.
3. Device structure of I-PSCs is not well considered and most of previous studies directly employed the device structure of OIH-PSCs. Future researches should pay more attention on designing device structure, such as choosing suitable HTM and ETM.
4. Physical insights into the working principle of I-PSCs are significantly lacking. Deep understanding of the carrier generation and lifetime, charge transfer and transport, ion behavior, charge accumulation, hysteresis, etc. should be obtained, which would help to improve the performance and stability of I-PSCs. As a promising alternative, halide double

PVKs were commonly proved to be an indirect bandgap semiconductor. Therefore, it will be important to devise strategies to engineer a direct bandgap.

Until now, several categories or structures of I-PVKs have been exploited, and it is still hard to comment which category will achieve the best performance. But the ABX_3 series has achieved significantly higher performance than other series, especially for $CsPbX_3$ perovskites. In addition, a number of studies have indicated that the ABX_3 series I-PVKs share some similar optoelectronic properties with their OIH-PVK counterparts. Therefore, it could be sure that ABX_3 series I-PVKs would still achieve the best performance in the near future. Besides, many promising double I-PVKs have been recently predicted, which may have great breakthrough after suitable synthetic routes are developed to effectively control film composition and morphology.

6.7 Conclusion

In this chapter, the recent progress in the studies of I-PSCs has been systematically reviewed. The crystal structures of I-PVKs were mainly categorized into ABX_3, A_2BX_6, $A_3B_2X_9$, and $A_2B^{1+}B^{3+}X_6$. Lead-based I-PVKs with the common structure of $CsPbX_3$ are the most studied materials, which have achieved PCEs over 10% with the highest value of 13.21%. The most efficient lead-free I-PVKs mainly have the crystal structure of $CsBX_3$ (B = Sn and Ge; X = Cl, Br, and I), and the PCEs of around 3% have been obtained by $CsSnX_3$. The perovskite-derived materials with the structures of A_2BX_6 and $A_3B_2X_9$ could be divided into Sn-based, Bi-based, and Sb-based materials, which only got very low PCEs of around 1%. Many promising double PVKs ($A_2B^{1+}B^{3+}X_6$) have been theoretically predicted. However, only $Cs_2AgBiBr_6$ (indirect bandgap = 1.9 eV) films have been prepared for PSCs, which obtained a PCE of over 2%. Though much progress has been made on I-PSCs, some important issues need to be addressed to further improve PCE and stability, including enhancing material stability, exploiting more synthesis methods, designing suitable device structure, and understanding the working mechanisms.

Questions

1. Schematically show the working principle and device structure of solar cells.
2. Describe the compositions and crystal structures of halide perovskites.
3. What types of halide perovskite compounds are there that have been used for building solar cells?
4. How do you classify lead-based inorganic perovskite solar cells?
5. What are the characteristics of lead-free inorganic perovskite solar cells?

References

Chang X., Li W., Zhu L., Liu H., Geng H., Xiang S., Liu J., and Chen H. (2016). *ACS Appl. Mater. Interfaces*, **8**, 33649.

Chen H. N., Xiang S. S., Li W. P., Liu H. C., Zhu L. Q., and Yang S. H. (2018). *Sol. RRL*, **2**(2), 1700188.

Chung I., Lee B., He J., Chang R. P. H., and Kanatzidis M. G. (2012). *Nature*, **485**, 486.

Eperon G. E., Paterno G. M., Sutton R. J., Zampetti A., Haghighirad A. A., Cacialli F., and Snaith H. J. (2015). *J. Mater. Chem. A*, **3**, 19688.

Giustino F. and Snaith H. J. (2016). *ACS Energy Lett.*, **1**, 1233.

Greul E., Petrus M. L., Binek A., Docampo P., and Bein T. J. (2017). *Mater. Chem. A*, **5**, 19972.

Harikesh P. C., Mulmudi H. K., Ghosh B., Goh T. W., Teng Y. T., Thirumal K., Lockrey M., Weber K., Koh T. M., Li S., Mhaisalkar S., and Mathews N. (2016). *Chem. Mater.*, **28**, 7496.

Kim Y., Yang Z., Jain A., Voznyy O., Kim G.-H., Liu M., Quan L. N., García de Arquer F. P., Comin R., Fan J. Z., and Sargent E. H. (2016). *Angew. Chem. Int. Ed.*, **55**, 9586.

Selig O., Sadhanala A., Müller C., Lovrincic R., Chen Z., Rezus Y. L. A., Frost J. M., Jansen T. L. C., and Bakulin A. A. (2017). *J. Am. Chem. Soc.*, **139**, 4068.

Chapter 7

Application in Electrocatalytic Water Splitting

Owing to the limited natural fossil fuel reserve and the current global environmental crisis, there has been a growing need for developing clean energy sources (e.g., solar, wind, and hydraulic power) as alternatives to fossil fuels, which are widely recognized as the only feasible option to ensure a sustainable development of the world economy and society. However, such clean energy sources can only generate electricity intermittently, posing a grand challenge in using them in an efficient and reliable way, and thus, it is of great significance to develop highly efficient and environmentally benign energy conversion and storage technologies. In order to ensure continuous energy supply on demand, electricity has to be somehow stored; an appealing way to solve this problem is the conversion of electricity to hydrogen (H_2). With the highest gravimetric energy density among the currently known or used fuels and a zero-carbon footprint, hydrogen has been considered a promising energy carrier to fulfill our need for future fuel applications. At present, as the main H_2 production route, the steam-reforming process from fossil resources not only aggravates the consumption of fossil fuels, but also contributes to the global CO_2 emission. Therefore, seeking a

---

Materials and Interfaces for Clean Energy
Shihe Yang and Yongfu Qiu
Copyright © 2022 Jenny Stanford Publishing Pte. Ltd.
ISBN 978-981-4877-66-4 (Hardcover), 978-1-003-14223-2 (eBook)
www.jennystanford.com

clean, renewable, and efficient scheme for H_2 production is urgently needed. Splitting water ($2H_2O \rightarrow O_2 + 2H_2$) is highly desirable in this regard, because an amount of 4.92 eV free energy can be stored in chemical bonds when two H_2O molecules are split into one O_2 and two H_2 molecules with no emission of greenhouse gases and other pollutant gases. The hydrogen evolution reaction (HER) is the cathodic half-reaction of water splitting, providing an ideal path to the "hydrogen-based economy." It is clear that Pt is still the best catalyst for the HER showing the highest exchange current density. Unfortunately, the widespread applications of Pt-based or Pt-group metal catalysts are hindered by their high cost, low earth-abundance, and insufficient stability.

The current challenge is to develop efficient, durable, and low-cost electrocatalytic materials for the HER. Attention is now shifted toward developing efficient and durable alternative HER electrocatalysts based on inexpensive and earth-abundant elements to replace noble metal catalysts. There have been remarkable advances in using transition-metal chalcogenides, carbides, nitrides, and phosphides as potential candidates to replace Pt due to their promising HER performance. However, the existing HER electrocatalysts based on non-noble metal compounds are still unsatisfactory in terms of both activity and stability. To further enhance the HER performance, very recently, researchers have put forward that binary-nonmetal atom incorporation with controllable disorder engineering and modification of the electronic structure may realize the synergistic modulations of both activity and conductivity for efficient HER performance. The incorporation of foreign nonmetal atoms into transition-metal compound (TMC) lattices to fabricate binary-nonmetal TMCs for HER electrocatalysis can shift the hydrogen adsorption free energy closer to thermal neutral by tuning the electronic structure and distorting the lattices of the parent TMCs. This strategy was proved by $MoS_{2(1-x)}Se_{2x}$ systems, which showed better HER catalytic performance than MoS_2 and $MoSe_2$. Similar phenomena were also observed for $WS_{2(1-x)}Se_{2x}$ when compared with WS_2 and WSe_2, $MoS_{0.94}P_{0.53}$ when compared with MoS_2 and MoP, and CoS|P when compared with CoS_2 and CoP. Along the same lines, P-modified WN/rGO also shows improved electrocatalytic performance toward the HER due to the enhanced interaction between WN and graphene with P.

Therefore, binary-nonmetal transition-metal compound (BNTMC) catalysts have emerged as alternative catalysts for hydrogen evolution. Remarkably, some of them have been found to exhibit comparable performance to Pt-containing catalysts for the HER. The phosphosulfide (MoP|S) catalyst exhibited high HER activities with a large exchange current density (0.57 mA/cm^2), high turnover frequency (TOF) value (0.75 s^{-1} at 150 mV), low overpotential (64 mV at the current density of 10 mA/cm^2), and superb durability (without activity decline even after 1000 operation cycles). Pyrite-type cobalt phosphosulfide (CoPS) electrodes achieved a catalytic current density of 10 mA/cm^2 at overpotentials as low as 48 mV. MoS$_{2(1-x)}$Se$_{2x}$/NiSe$_2$ hybrids exhibited excellent HER performance with a low overpotential of 69 mV at the current density of 10 mA/cm^2, a high exchange current density of 0.3 mA/cm^2, a turnover frequency of 0.22 s^{-1} at 150 mV, and no obvious activity degradation after more than 16 h of operation at high potentials. The path-breaking studies over the past 3 years have ushered in a golden era of binary-nonmetal TMCs in HER electrocatalysis.

Electrochemical (EC) water splitting separates the whole process into two half-reactions: hydrogen evolution reactions (HERs) and oxygen evolution reactions (OERs). As is well known, compared to HER, OER is much more kinetically sluggish and requires a very high overpotential, thus limiting the water-splitting efficiency and hindering the development of the hydrogen production industry based on water splitting.

The state-of-the-art catalysts of RuO$_2$ and IrO$_2$ show effective water oxidation activity. However, the high cost of Ir ($16,181 per kg) and Ru ($2000 per kg) significantly limits their large-scale utilization. Fortunately, the first-row transition-metal-based compounds with controlled chemical composition and microstructures have been found to have comparable water-splitting performance, with much more bountiful resources and lower prices. Among these materials, transition-metal-based layered double hydroxides (TM LDHs), with the general formula of $[M_{1-x}^{2+}M^{3+}_x(OH)_2]^{x+}(A^{n-})_{x/n} \cdot mH_2O$, consisting of edge-sharing hydroxyl coordinated octahedrons and intercalated anions between hydroxyl layers, have been extensively studied, which surprisingly have exhibited even higher OER activities with overpotentials as low as ~200 mV, which are smaller than that of RuO_2 and IrO_2 (~250 mV). In spite of the tremendous research progress that has been made in developing TM LDH as efficient OER catalysts, the understanding of the underlying reasons for the high water oxidation performance of TM LDH is still limited. For instance, the synergistic effect between the divalent and trivalent cations and their effects on OER remain elusive.

In this chapter, a brief introduction of HER fundamentals is started. This is followed by the summary of some main approaches of frequently used synthetic strategies to fabricate nanostructured binary-nonmetal TMCs as a representative class of materials for the HER. Then, the relationships between the tuned physicochemical structure and the performance of binary-nonmetal TMCs in the HER are discussed. And then, the recent developments of transition-metal-based layered double hydroxides (TM LDHs) with unary, binary, and ternary transition-metal ions and their OER catalytic performances are surveyed, focusing on the effects of metal combinations that have strong influences on the OER activity. In the final section, the prospects and the major challenges for the binary-nonmetal TMC and TM LDHs catalyst materials are presented.

7.1 HER Fundamentals

Before we turn to binary-nonmetal TMCs in HER electrocatalysis, it is instructive to briefly touch upon some fundamentals of the HER. The mechanism of the HER is commonly treated as a combination of two out of three elementary steps: Volmer–Tafel or Volmer–Heyrovsky, including the electrochemical adsorption of H^+, chemical desorption of H_2, and electrochemical desorption of H_2. Volmer step (discharge step):

$$H^+ + e^- \rightarrow H_{ad} \qquad (7.1)$$

Heyrovsky step (electrochemical atom + ion reaction step):

$$H_{ad} + H^+ + e^- \rightarrow H_2 \qquad (7.2)$$

Tafel step (atomic combination step):

$$2H_{ad} \rightarrow H_2 \quad (7.3)$$

For electrocatalysis reactions, the activities of catalysts depend sensitively on the energetics of the interactions between the metal surfaces and the key reaction intermediates, including adsorption, bond making, bond breaking, and desorption. As indicated by reaction mechanisms (7.1), (7.2), or (7.1)–(7.3), a good HER catalyst should form a sufficiently strong bond with adsorbed H_{ad} for facilitating the electron-transfer process, but at the same time should not be too strong to ensure a facile surface bond breaking for releasing the gaseous H_2 product (Morales-Guio et al., 2014).

7.1.1 Hydrogen Adsorption on Catalyst Surface

HER electrocatalysis is known to be strongly correlated with the chemisorption free energy of atomic hydrogen (ΔG_H) on the catalyst surface, which is optimal at 0 eV because the hydrogen adsorption on the catalyst surface should be neither too weak nor too strong. The volcano plot (Fig. 7.1) shows the logarithmic exchange current density of transition metals as a function of the chemisorption free energy of atomic hydrogen.

According to the Sabatier principle, neither too strong nor too weak binding would favor the overall reaction because strong or weak binding leads to either poor adsorption of reactants or difficulty in removing final products. As compared with the Pt-group metals, the earth-abundant, non-noble Ni, Fe, Co, Mo, and W metals are all located on the left branch of the volcano curve, indicating sluggish HER kinetics on these metal surfaces due to the strong binding of hydrogen. Previous studies have discovered that the inclusion of some moderately electronegative nonmetal atoms (e.g., C, N, P, and S) could modify the nature of the d-band of parent metals, affording characteristics resembling those of Pt.

Density functional theory (DFT) calculations have established hybridization between the metal d-orbitals and the carbon (nitrogen) sp-orbitals in transition-metal carbides (nitrides). Because of the tensile strain induced in the carbide and nitride surfaces upon incorporation of C or N atoms into the metal lattice, the metal lattice expands and the metal–metal distance increases, and as a result, hydrogen adsorbs strongly on those metal sites. DFT calculations have demonstrated that hydrogen adsorbed more strongly on metal-terminated carbide surfaces while more weakly on carbon-terminated carbide surfaces than on the corresponding closest-packed pure metal surfaces. More recently, Peterson et al. found that although hydrogen adsorbed strongly on metal-terminated carbide surfaces, when increasing hydrogen coverage, hydrogen binding on metal carbide surfaces became threefold weaker than that on the parent metal surfaces. Therefore, they considered that the increased HER activities of metal carbides could be attributed to the higher sensitivity to coverage-induced weakening of hydrogen-binding strength. Besides the strain-induced d-band modification, the ligand effect due to the electron transfer from metal to carbon or nitrogen would lead to a deficiency in the d-band occupation near the Fermi level in comparison with the parent metal. Due to the deficiency in the d-band occupation, carbide/nitride surfaces could have a reduced ability to donate d electrons to hydrogen, and thus a weakened hydrogen-binding strength.

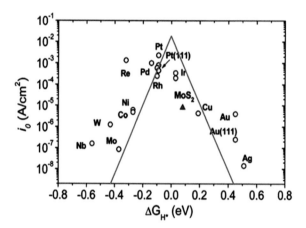

Figure 7.1 A volcano plot of experimentally measured exchange current density as a function of the DFT-calculated Gibbs free energy of hydrogen adsorption. The simple kinetic model proposed by Nørskov et al. to explain the origin of the volcano plot is shown as the solid lines. (Hu et al., 2017; Zheng et al., 2015).

The physicochemical properties of transition-metal phosphides are similar to those of transition-metal carbides/nitrides. However, the differences in the electronegativity and atomic radius between phosphorus and carbon/nitrogen bring some different properties. Different from metal carbides and nitrides where carbon and nitrogen atoms nest in the interstitial spaces of metal atom lattices to form relatively simple crystal structures (e.g., body-centered cubic crystal structures for molybdenum nitrides or hexagonal close-packed structures for molybdenum carbides), the crystal structure of metal phosphides is based on trigonal prisms due to the large radius of phosphorus atoms. Rather than a layered structure observed in metal disulfides, metal phosphides tend to form an isotropic crystal structure, which benefits greater access of H to active sites on the crystallite surfaces. The electronegativity of phosphorous is low in comparison with those of carbon and nitrogen, producing a weak ligand effect in metal phosphides. In the case of Ni_2P, the Ni hollow sites of Ni_2P are strong hydride acceptors and the first hydrogen adsorption on the $Ni_2P(001)$ surface is only slightly weaker than that on the metal Ni(111) surface due to the weak ligand effect. The strong hydrogen-binding strength on the Ni_2P surface seems contradictory to the high activity of Ni_2P for the HER. However, a theoretical study from Liu et al. demonstrated that strong H–Ni interaction on the $Ni_2P(001)$ surface could give rise to a "H-poisoning effect" that allowed more H atoms to moderately bond on the Ni–P bridge sites after Ni hollow sites were occupied, and thus made the HER activity of the Ni_2P catalyst comparable to those of Pt and [NiFe] hydrogenase. A similar H-poisoning effect also exists in the case of MoP. Wang et al. found that the free energy of hydrogen adsorption on MoP(001) P-terminated planes could approach 0 eV at the H coverage of 50% to 75%.

The free energy of hydrogen adsorption on layered structured transition-metal chalcogenides is dependent sensitively on the materials' facets. Taking MoS_2 as an example, the orbitals of HOMO states are mainly localized at the S sites at the edge, which means that the localized electrons are mainly at the edge of MoS_2 for charge exchange of protons. DFT calculations in terms of the HER free-energy diagram have elucidated that the MoS$_2$ plane sites are inert (ΔG_H is ~1.9 eV), while edge sites especially the [1010] Mo edges are active for the HER with a ΔG_H of ~0.08 eV, which is even closer to 0 eV than that of the state-of-the-art Pt surfaces (ΔG_H is ~0.1 eV). Motivated by this understanding, extensive efforts have been devoted to developing MoS$_2$ nanostructures to maximize the active edge sites, including nanoparticles, nanowires, amorphous films, films with ordered double-gyroid networks, nanosheets, and defect-rich nanosheets.

Optimized hydrogen adsorption on the planar sites has been observed for exfoliated MoS$_2$ layers after converting the 2H phase to the 1T phase (Voiry et al., 2013). Generally, for the 2H phase, the d-orbital of Mo splits into three degenerate orbitals d_{z^2}, $d_{x^2-y^2,xy}$, and d_{xy-yz} with an energy gap of 1 eV between the d_{z^2}- and $d_{x^2-y^2,xy}$-orbitals, while for the tetragonal symmetry of the 1T phase, the d-orbital splits into two degenerate states $d_{xy,yz,zx}$- and $d_{x^2-y^2,z^2}$-orbitals (Fig. 7.2A). Thus, up to six electrons can fill the $d_{xy,yz,zx}$-orbital in the 1T phase, while only two electrons can fill the d_{z^2}-orbital in the 2H phase. Because the p-orbitals of S are located at a much lower energy than the Fermi level, the filling of d-orbitals determines the nature of the different phases in MoS$_2$ compounds. Complete filling of d-orbitals in transition-metal chalcogenides results in semiconducting behavior (2H), while partial filling gives rise to metallic behavior (1T). For example, 1T MoS$_2$ or 1T WS$_2$ thin films exhibit a conductivity of ca. 0–100 S/cm compared with 10^{-5} S/cm of their 2H phase counterparts. More importantly, it should be noted that the HER performance of 1T MoS$_2$ remains unaffected after edge oxidation (Fig. 7.2B), indicating that edges are not the only active sites in the 1T MoS$_2$ nanosheets. Later on, Tsai et al. have calculated the free energy of hydrogen adsorption to estimate the catalytic activity and stability of the edges and basal planes of various transition-metal chalcogenides. Their results revealed that the basal planes of transition-metal chalcogenides can be catalytically activated by optimizing the free energy of hydrogen adsorption. The ΔG_H of MoS$_2$ on the basal planes is significantly decreased in the case of the metallic 1T phase (0.12 eV) compared with that of the semiconducting 2H phase (1.9 eV).

HER Fundamentals | 197

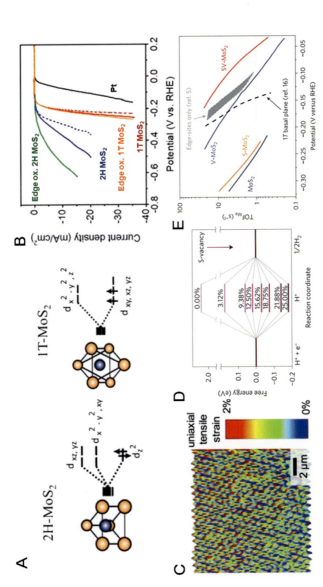

Figure 7.2 (A) The crystal-field-splitting-induced electronic configuration of 2H and 1T MoS$_2$, where orange spheres are S atoms, and the blue–gray sphere is the Mo atom. (B) Polarization curves of 1T and 2H MoS$_2$ nanosheet electrodes before and after edge oxidation. iR-Corrected polarization curves of 1T and 2H MoS$_2$ are shown in dashed lines. (C) Color-coded strain distribution in strain-textured monolayer MoS$_2$ (14 × 14 μm^2) obtained from scanning micro-Raman spectroscopy. The peak strain magnitude on the top of nanopillars is about 2% (uniaxial tensile strain). (D) Free energy of hydrogen adsorption versus the reaction coordinate of the HER for the S-vacancy range of 0–25%. (E) TOF$_{Mo}$ of various MoS$_2$ samples. (Hu et al., 2017; Li et al., 2016).

In addition, Chhowalla et al. demonstrated that exfoliated WS$_2$ nanosheets became distorted when converting the 2H phase to the 1T phase and the distortion induced a strain on the WS$_2$ nanosheets. DFT calculations revealed that the strain in the 1T WS$_2$ nanosheets led to the enhancement in the density of states near the Fermi level, which facilitated hydrogen binding, and a near-zero free energy of hydrogen adsorption was achieved with a strained value of 2.7%. More recently, Zheng et al. and Nørskov et al. have demonstrated that in the case of 2H MoS$_2$, strain could only slightly increase the HER activity, while the combination of strain and S-vacancies could significantly increase the HER activity. They transferred MoS$_2$ monolayers onto a SiO$_2$ nanopillar array substrate capped by an Au/Ti film, so that the MoS$_2$ monolayers are distorted with a larger strain on top of the nanopillars and a smaller strain between nanopillars (Fig. 7.2C). Experimental and theoretical results indicated that S-vacancies acted as new highly active and tunable catalytic sites for the HER and the strained MoS$_2$ with S-vacancies (SV–MoS$_2$) showed an enhanced HER activity significantly influencing the density of states near the Fermi level, which helped to lower the free energy of hydrogen adsorption (Fig. 7.2D). The turnover frequency per surface Mo atom (TOF$_{Mo}$) of the SV–MoS$_2$ catalysts is even higher than those of the MoS$_2$ edge sites and the basal plane of 1T phase MoS$_2$ (Fig. 7.2E).

7.1.2 Reaction Pathways

Besides the free energy of hydrogen adsorption, HER kinetics are strongly dependent on the reaction pathway, which in turn can be influenced by the catalyst, applied potential, etc. The HER current is typically expressed by the Butler–Volmer equation,

$$\eta = \frac{2.303RT}{\alpha nF}\log i_k - \frac{2.303RT}{\alpha nF}\log i_0 \qquad (7.4)$$

where α is the transfer coefficient, and i_0 is the exchange current density.

At large overpotentials (the linear part for Tafel fitting for the HER), $e^{\frac{\alpha nF\eta}{RT}} \ll e^{\frac{-(1-\alpha)nF\eta}{RT}}$. Thus, $e^{\frac{-(1-\alpha)nF\eta}{RT}}$ in Eq. (7.4) becomes negligible and Eq. (7.4) becomes

$$i_k = i_0 \times e^{\frac{\alpha n F \eta}{RT}} \quad (7.5)$$

or

$$\eta_{\text{diff}} = \frac{2.303RT}{\alpha n F} \log i_{k,d} - \frac{2.303RT}{\alpha n F} \log i_0 \quad (7.6)$$

From Eq. (7.6) we find that the aforementioned kinetic treatment naturally yields a relation of the Tafel form, which specifies the linear relationship between the overpotential (η) and log i_k, with a slope $\left(\frac{2.303RT}{\alpha n F}\right)$ termed the Tafel slope. Therefore, a small Tafel slope corresponds to a steep rise in the electrocatalytic current density with increasing overpotential.

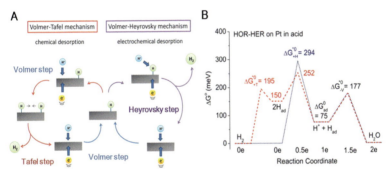

Figure 7.3 (A) The mechanism of hydrogen evolution on the surface of an electrode in acidic solution. (B) Free-energy diagram of the HOR and HER at 0 V on Pt in acidic solution, constructed using fitted parameters with the Volmer–Tafel pathway (red line) and the Volmer–Heyrovsky pathway (blue line). (Hu et al., 2017; Elbert et al., 2015).

The possible HER reaction pathways are based on the three principal reactions (Volmer–Tafel or Volmer–Heyrovsky pathways, as shown in Fig. 7.3A) on different types of surfaces, including metals, transition-metal compounds, and nonmetallic materials. Assuming that one of the reactions in each pathway is the rate-determining step (rds), it leads to four limiting cases: (i) Volmer–Tafel (rds); (ii) Volmer (rds)-Tafel; (iii) Volmer–Heyrovsky (rds); and (iv) Volmer (rds)-Heyrovsky. The Tafel slope is a common parameter that can be used to discern the possible reaction pathway and the rate-determining step of the HER. If the discharge reaction

(Volmer step) is fast, the reactant mass transport is coupled to an atomic combination reaction (Tafel step), which together becomes the rate-determining step. Assuming that the Nernstian diffusion-controlled HER obeys the Tafel form, we have

$$\eta_{\text{diff}} = \frac{2.303RT}{\alpha nF}\log i_{k,d} - \frac{2.303RT}{\alpha nF}\log i_0 \quad (7.7)$$

Thus, a Tafel plot of η_{diff} versus $\log(i_{k,d})$ yields a slope of $2.303RT/\alpha nF$, and the Nernstian diffusion-controlled HER should have the lowest Tafel slope of $2.3RT/2F$, that is, 29 mV/dec at 293 K. Empirically, the Tafel slope value within the range from 29 to 38 mV/dec is attributed to the slow atomic combination to form H_2 (Tafel step), for example, on the Pt (100) surface in an acidic electrolyte. The higher Tafel slopes of electrocatalysts indicate that the charge transfer step (Heyrovsky or Volmer step) is the rate-determining step. For the case of transition-metal compound catalysts, the Tafel slopes have been observed in a wide range from 40 to 110 mV/dec, suggesting the Volmer–Heyrovsky pathway. If the Volmer step is fast and is followed by a rate-determining Heyrovsky step (Volmer–Heyrovsky (rds) pathway), then $n = 2$ and $\alpha = 3/4$, leading to a Tafel slope of 38 mV/dec at 293 K. If the reaction pathway shows a slow discharge mechanism, the Tafel slope should be above 116 mV/dec, indicating that $n = 2$ and $\alpha = 1/4$. When the rates of the Volmer step and Heyrovsky step become comparable, the Tafel slope should be intermediate between 38 and 116 mV/dec. However, it is worth noting that these values are based on a strict set of assumptions. Many factors may lead to their deviation, for example, the surface coverage of hydrogen might be intermediate and potential dependent. Conway et al. showed that Tafel slope values for the HER of Hg and Ni, determined over a wide range of temperature, were not simply $2.3RT/\alpha nF$ but could be independent of T. Moreover, it is hard to distinguish at present between pathways (ii) and (iv) from the Tafel slope because for both pathways, the rate-determining reaction is the Volmer step. Multiple pathways can also occur in parallel with one another.

The reaction barriers between two states can significantly influence the overall reaction rate and can be calculated by obtaining the minimum energy pathway. Adzic et al. developed an HOR/HER kinetic analysis from a dual-pathway model to describe the complex kinetic behavior of catalysts' surfaces over the whole relevant potential region. Similar to many chemical reactions, HER processes have to surmount certain activation energy barriers to progress to completion. Figure 7.3B illustrates a free-energy diagram constructed using four fitted standard free energies (($\Delta G_{+H}^{*0} \Delta G_{-V}^{*0}$), ΔG_{ad}^{0}, ΔG_{+H}^{*0}, ΔG_{-V}^{*0}) for visualizing the reaction barriers for the HOR (positive direction) and HER (negative direction) in Volmer–Tafel or Volmer–Heyrovsky reaction pathways. It can be seen that the activation barrier at 0 V for the Tafel reaction is lower than that for the Heyrovsky reaction demonstrating that the Volmer–Tafel pathway dominates at small overpotentials and the Tafel step is the rate-determining step ($\Delta G_{+H}^{*0} > \Delta G_{-V}^{*0}$) on the Pt surface, which is consistent with the results deduced from Tafel slopes. The reaction barriers are correlated with the electrode potential. Nørskov et al. developed a two-step extrapolation scheme that could derive the barrier values under different electrode potentials. The DFT calculations show that the chemical potential μ is related to the electrode potential U via the following simple equation: $\mu = -eU$, where e is the transferred charge. It is obvious from reactions (7.1) and (7.2) that the Volmer state (2H$^+$ + 2e$^-$) changes by $2U$ from the H$_2$ state, while the Heyrovsky state (1 H$^+$ + 1e$^-$ +H*) changes by $1U$ plus the H adsorption energy, indicating a potential-dependent minimum energy pathway. More details were explicitly elaborated in the literature.

As argued above by drawing on previous studies, introducing nonmetal atoms into transition-metal lattices to form transition-metal compounds (TMCs) can significantly modify the d-band of parent metals and shift the hydrogen adsorption free energy closer to thermal neutral. However, there is still much room for improvement in terms of the HER activity. Theoretical calculations have indicated that further incorporation of foreign nonmetal atoms into TMC lattices to fabricate binary-nonmetal TMCs (BN-TMCs) for HER catalysis can shift the hydrogen adsorption free energy even closer to thermal neutral by tuning the electronic structure and bandgap and distorting the lattices of the parent TMCs. The efforts along this line of research have indeed resulted in superior catalytic performance for the HER, affording characteristics quite resembling those of Pt. This is actually the main topic reviewed here. In the following sections, the relationships between the tuned physicochemical structure and the performance of the binary-nonmetal TMCs in the HER will be presented and discussed in detail.

7.2 Binary-Nonmetal TMCs for Hydrogen Evolution

In the designing process of robust electrocatalysts, the availability of the volcano plot for catalysts with different adsorption strengths of adsorbates in each reaction can be used to predict the best performance of each catalyst for specific reactions, as well as to achieve the optimal adsorption energy of intermediates. Introducing foreign nonmetal atoms into transition-metal compound lattices can cause significant tuning of the electronic structure of the parent TMCs and shift the hydrogen adsorption free energy closer to thermal neutral and, thus, result in superior catalytic performance for the HER. Furthermore, incorporating bigger or smaller radii of foreign nonmetal atoms can cause a distortion of TMC lattices, resulting in favorable bond breaking of the adsorbed molecules and thus increasing active sites and improving HER performance. Moreover, foreign nonmetal atom incorporation can lead to a smaller bandgap, and thus better conductivity compared to that of the parent TMCs, and easy electrocatalysis. Therefore, binary-nonmetal TMCs have gained considerable interest for use as HER electrocatalysts very recently.

7.2.1 Transition-Metal Sulfoselenides (MSSe)

Foreign atom incorporation into TMCs is a technique that can alter their structure and properties depending on the dopant (if the dopant is well matched in terms of size, valence, and coordination). Due to the similarities of S and Se, one of them could be doped into compounds formed by the other one without phase separation. First-principles calculations on S-doped $SnSe_2$ systems have been used to predict the potentially useful electronic and optical properties that may arise from S inclusion and found that $SnSe_{2(1-x)}S_{2x}$ alloys could be favorably formed by S substitution for Se atoms in $SnSe_2$ monolayers because of the negative forming energies. Experimental studies have shown the successful synthesis of $SnSe_{2(1-x)}S_{2x}$ alloy nanosheets with tunable chemical compositions and proved that the bandgaps of the SnSe$_{2(1-x)}$S$_{2x}$ alloy nanosheets can be discretely modulated from 2.23 to 1.29 eV with increasing Se content. Theoretical advances by Komsa et al. have indicated that selenium atom doping may modulate the electronic structure and band structure of MoS$_2$ nanosheets. By applying the effective band structure approach, it can be seen from Fig. 7.4 that the general features of the band structures of the MoS$_{2(1-x)}$Se$_{2x}$ alloy are similar to those of their binary constituents (MoS$_2$ and MoSe$_2$). Most importantly, the short-range atomic order favors having dissimilar atoms in the nearest neighbor sites in chalcogen sublattices, indicating that the incorporation of selenium heteroatoms into MoS$_2$ lattices is a facile thermolytic process. Nørskov et al. used DFT calculations to determine the free energies of hydrogen adsorption on the active edge sites of MoS$_2$, MoSe$_2$, WS$_2$, and WSe$_2$. They found that the hydrogen adsorption is slightly weak on the sulfided Mo edges of MoS$_2$ (ΔG_H is ~0.08 eV) and selenided W-edges of WSe$_2$ (ΔG_H is ~0.17 eV), while slightly strong on the selenided Mo edges of MoSe$_2$ (ΔG_H is ~0.04 eV) and sulfided W-edges of WS$_2$ (ΔG_H is ~0.04 eV). So it is reasonable to achieve a thermoneutral ΔG_H by designing binary-nonmetal MoS$_{2(1-x)}$Se$_{2x}$ and WS$_{2(1-x)}$Se$_{2x}$ alloys for hydrogen evolution, and as a result, higher HER activities can be achieved by changing the S/Se ratio.

Ultrathin S-doped MoSe$_2$ nanosheets possessing higher HER activity compared to pristine MoSe$_2$ were first developed by Yan et al., and they exhibit highly active HER electrocatalysis with an overpotential of 156 mV at a current density of 30 mA/cm^2 and a small Tafel slope of 58 mV/dec. This study demonstrated the improvement of HER activity by substituting S atoms into MoSe$_2$ compounds. Sampath's group reported that few-layered molybdenum sulphoselenide (MoS$_{2(1-x)}$Se$_{2x}$) nanosheets possess higher HER activity compared to pristine few-layered MoS$_2$ and MoSe$_2$, and the highest HER activity was found when the molar ratio of S and Se was 1 : 1. This study suggested that the electrocatalytic activity could be easily tuned by adjusting the composition of target materials, that is, the tunable electronic structure of MoS$_2$/MoSe$_2$ upon Se/S incorporation probably assists in the realization of high HER activities. Xue et al. synthesized hierarchical ultrathin molybdenum sulphoselenide MoS$_{2(1-x)}$Se$_{2x}$ nanosheets by a one-step hydrothermal method with Se powder and thioacetamide

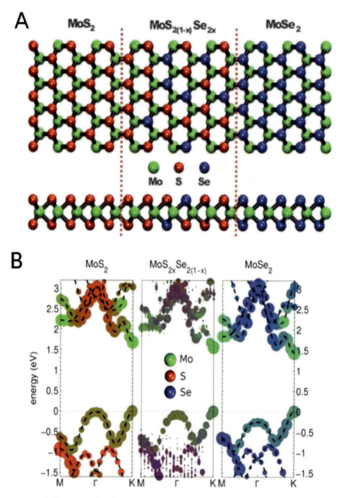

Figure 7.4 (A) Top and side views of the atomic structures of MoS$_2$ (left), the MoS$_{2(1-x)}$Se$_{2x}$ alloy (middle), and MoSe$_2$ (right). (B) The effective band structures of MoS$_2$, MoS$_{1.2}$Se$_{0.8}$ alloy, and MoSe$_2$.

as the Se-precursor and S-precursor, respectively. By adjusting the S/Se-precursor ratio, the authors achieved good control over the composition and chemical structure of the final products. The hierarchical MoS$_{2(1-x)}$Se$_{2x}$ nanosheets exhibit excellent composition-dependent ferromagnetism and electrocatalytic properties for the HER, and MoS$_{0.98}$Se$_{1.02}$ nanosheets possess the highest HER activity with the Tafel slope of 57 mV/dec and an overpotential of 271 mV to reach a current density of 10 mA/cm². Atomically thin uniform alloy MoS$_{2(1-x)}$Se$_{2x}$ triangular nanosheets have been simultaneously synthesized with complete composition (0 ≤ x ≤1) tunability using a very simple one-step temperature gradient assisted chemical deposition approach by Duan et al. The nanosheets show a fully tunable chemical composition and optical properties, suggesting potential applications in functional nanoelectronic and optoelectronic devices. Ultrathin alloy molybdenum sulphoselenide MoS$_{2(1-x)}$Se$_{2x}$ nanoflakes with monolayer or few-layer thicknesses were prepared by Liu et al. through a high-temperature solution method. It is observed that successful preparation of MoS$_{2(1-x)}$Se$_{2x}$ alloys as x increases from 0 to 1 with a fully tunable chemical composition is free of phase separation. The authors corroborate that the introduction of selenium continuously modulates the d-band electronic structure of molybdenum, which tunes hydrogen adsorption free energy and consequently electrocatalytic activity.

Like molybdenum sulphoselenides (MoS$_{2(1-x)}$Se$_{2x}$), the formation of tungsten sulphoselenides (WS$_{2(1-x)}$Se$_{2x}$) modifies the nature of the d-band electronic structure of tungsten giving rise to a thermoneutral hydrogen adsorption free energy and high HER activity. Because the p-orbitals of S are located at a much lower energy than the Fermi level, the filling of the d-orbitals of metal determines the nature of WS$_2$ compounds. In general, the valence band maximum (VBM) of WS$_2$ mainly consists of the d_{z^2}-orbital and the conduction band minimum (CBM) mainly consists of the $d_{x^2-y^2, xy}$- orbitals, and the bandgap is 1.58 eV for monolayer WS$_2$. Although the effect of chalcogen atoms on the electronic structure is minor compared with that of metal atoms, the interaction between W and chalcogen atoms is an important issue for the d-band structure. As Se atoms replace S atoms in the WS$_{2(1-x)}$Se$_{2x}$ alloy, the bonding between W and Se becomes more covalent due to the lower electronegativity value of Se atoms compared with that of S atoms, resulting in the broadening of the fully occupied d$_z$-band. Therefore, the tunable bandgap of monolayer WS$_{2(1-x)}$Se$_{2x}$ was achieved by varying the ratio of Se and S. Xiang et al. reported the first realization of a tunable bandgap in monolayer WS$_{2(1-x)}$Se$_{2x}$ triangular domains using a chemical vapor deposition route, which is achieved by varying the Se content. The monolayer WS$_{2(1-x)}$Se$_{2x}$ alloy exhibits a lower Tafel slope of 85 mV/dec than monolayer WSe$_2$ (100 mV/dec) and WS$_2$

(95 mV/dec), and the lowest overpotential (≈80 mV) and the largest exchange current density as well. This study also found that the non-ohmic contact between the samples and GC electrode due to the residual polymethyl methacrylate on the surface of the samples could be responsible for the larger Tafel slope compared to that of WS_2 reported by others.

Direct growth of active metal sulphoselenides on conductive substrates such as carbon fiber, 3D graphene, metal foam, etc., will be favorable for electron transfer from electrode materials to current collectors, accelerating the HER process. The key challenge is to find suitable supports with a high surface area, high porosity, and good conductivity. He et al. prepared $WS_{2(1-x)}Se_{2x}$ nanotubes with controllable S and Se amounts on carbon fiber using the chemical vapor deposition method. The as-grown $WS_{0.96}Se_{1.04}$ nanotubes provide both an increased number of active sites and higher conductivity and exhibit improved electrocatalytic properties for hydrogen evolution with a lower overpotential (260 mV at the current density of 10 mA/cm^2), higher exchange current density (0.029 mA/cm^2), and smaller charge transfer resistance (204 Ω at the overpotential of 128 mV) compared with WS_2 and WSe_2. Ren et al. reported a simple and efficient strategy to synthesize a highly active $MoS_{2(1-x)}Se_{2x}/NiSe_2$ hybrid for hydrogen evolution. In their strategy, direct selenization of Ni foam in an Ar atmosphere was first performed to convert the Ni foam to porous $NiSe_2$ foam and then directly grew binary-nonmetal $MoS_{2(1-x)}Se_{2x}$ particles on metallic $NiSe_2$ foam (Fig. 7.5A,B). Owing to the excellent electrical conductivity, porous structures, and high surface area of the $NiSe_2$ foam substrate, the $MoS_{2(1-x)}Se_{2x}/NiSe_2$ hybrid possesses an ultrahigh electrochemical surface area with the double-layer capacitance of 319 mF/cm^2, and a low Tafel slope of 42 mV/dec. DFT calculations found, as shown in Fig. 7.5C, that the hydrogen adsorption free energy on $MoS_{2(1-x)}Se_{2x}/NiSe_2$ (100) and $MoS_{2(1-x)}Se_{2x}/NiSe_2$ (110) decreased to 2.7 and 2.1 kcal/mol, which are much lower than those of $MoS_{2(1-x)}Se_{2x}/MoS_{2(1-x)}Se_{2x}$ (8.4 kcal/mol) and MoS_2/MoS_2 (10.6 kcal/mol). Electrochemical measurements showed that the $MoS_{2(1-x)}Se_{2x}/NiSe_2$ hybrid exhibited excellent HER performance with a low overpotential of 69 mV at the current density of 10 mA/cm^2, a high exchange current density of 0.3 mA/cm^2, and a turnover frequency of 0.22 s^{-1} at 150 mV (Fig. 7.5D). Following the success of the

Binary-Nonmetal TMCs for Hydrogen Evolution | 207

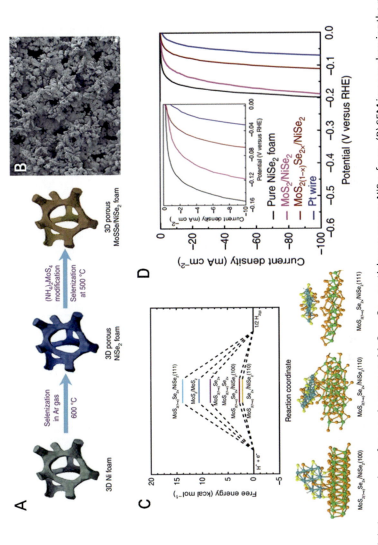

Figure 7.5 (A) The procedures for growing ternary $MoS_{2(1-x)}Se_{2x}$ particles on porous $NiSe_2$ foam. (B) SEM images showing the morphologies of ternary $MoS_{2(1-x)}Se_{2x}$ particles distributed on porous $NiSe_2$ foam. (C) Calculated free-energy diagram of hydrogen adsorption at the equilibrium potential for the $MoS_{2(1-x)}Se_{2x}/NiSe_2$ hybrid, and MoS_2 and $MoS_{2(1-x)}Se_{2x}$ catalysts. (D) The polarization curves recorded on $MoS_{2(1-x)}Se_{2x}/NiSe_2$ hybrid, $MoS_2/NiSe_2$ hybrid, and pure $NiSe_2$ foam electrodes compared with that of a Pt wire in 0.5 M H_2SO_4 solution.

MoS$_{2(1-x)}$Se$_{2x}$/NiSe$_2$ foam hybrid, Ren et al. has extended their attention to the WS$_{2(1-x)}$Se$_{2x}$/NiSe$_2$ hybrid using the same method. WS$_{2(1-x)}$Se$_{2x}$ is isostructural to MoS$_{2(1-x)}$Se$_{2x}$, and thus it is not surprising that the WS$_{2(1-x)}$Se$_{2x}$/NiSe$_2$ hybrid possesses outstanding catalytic performance with a large exchange current density (0.2 mA/cm^2), low overpotential (88 mV at the current density of 10 mA/cm^2), low Tafel slope (46.7 mV/dec), and good stability, which is better than that of WS$_2$ and WSe$_2$ catalysts.

Besides the adjustment of the electronic structure of MoS$_2$/MoSe$_2$ upon Se/S incorporation, the enhanced HER performance of the alloy can also be attributed to the different atomic radii of S and Se. In the field of catalysis, doping of TMCs can increase the density of catalytic sites, either by altering the morphology to expose more sites or by the creation of additional sites, especially in the basal plane of layers. The introduction of Se into MoS$_2$/WS$_2$ or S into MoSe$_2$/WSe$_2$ lattices will generate a slight distortion in the basal plane of the pristine crystal structure due to the larger radius of Se atoms compared with that of S atoms, which can bring about polarized electric field localized in the basal planes, resulting in favorable bond breaking of the adsorbed molecules. Hu et al. prepared component-controllable 3D vertically oriented MoS$_{2(1-x)}$Se$_{2x}$ alloy nanosheets directly on conductive carbon cloth. The S and Se percentage of the as-grown ternary MoS$_{2(1-x)}$Se$_{2x}$ alloy nanosheets can be easily controlled by altering the ratios of S- and Se-precursors. The basal plane of the as-grown vertically oriented MoS$_{2(1-x)}$Se$_{2x}$ alloy nanosheets in their study was demonstrated to be catalytically active by DFT calculations, which shows that the hydrogen adsorption free energy decreases first and then increases with increasing the Se content (x value) in the MoS$_{2(1-x)}$Se$_{2x}$ alloy and a better HER performance can be achieved with the x value in the range between 0.33 and 0.67.

At the end of this subsection, we want to add other nonmetal element incorporated transition-metal sulfides, which have also been investigated for the highly electrocatalytic HER. Oxygen-incorporated MoS$_2$ ultrathin nanosheets have been developed by Xie et al. to generate surface defects. As the degree of disorder increases, more unsaturated sulfur atoms are generated as active sites for HER catalysis, resulting in the improvement in electrochemical hydrogen evolution by activation of basal planes. Meanwhile, oxygen incorporation effectively reduces the bandgap of the MoS$_2$ catalyst, leading to the enhancement of the intrinsic conductivity. In addition, it was believed that such an enhancement of the HER performance of the O-doped MoS$_2$ catalyst could be attributed to not only the enhanced intrinsic conductivity and more active edge sites from the defect-rich structure, but also the expanded interlayer spaces, which improved the intrinsic conductivity of MoS$_2$ sheets with oxygen incorporation. Wang et al. constructed fullerene-like nickel oxysulfide hollow nanospheres by in situ growth on the surface of nickel foam. The NiOS/Ni foam composite exhibited high HER and OER performances simultaneously in 1 M KOH solution. Recently, Jin et al. deposited an amorphous MoS$_x$Cl$_y$ electrocatalyst on conducting vertical graphene by a temperature-controlled chemical vapor deposition process. The alloying of the nonmetal element Cl modifies the electronic structure of amorphous MoS$_x$ and introduces defect states within the bandgap. Moreover, the amorphous structures generally present more disorder and active sites for catalysis. The addition of Cl to amorphous MoS$_x$ would make it more disordered to form more active sites for the HER, as confirmed by the larger double-layer capacitance (Cdl) as compared with that of crystalline MoS$_2$. Therefore, simultaneous disorder engineering and electronic structure tailoring by chlorine incorporation into MoS$_x$ become an effective solution to improve the catalytic activity for the HER. Sun et al. synthesized N-doped WS$_2$ nanosheets by a one-step sol–gel process. According to the DFT calculations, they found that the p-orbitals of the N atom are strongly hybridized with the d-orbitals of the nearest neighbor W atoms and the p-orbitals of S atoms at the Fermi level in N-doped WS$_2$, resulting in that the N-doped WS$_2$ monolayer possesses a narrower bandgap of 1.5 eV than the pristine WS$_2$ monolayer (1.8 eV) and an improved intrinsic conductivity. The overpotential for this catalyst to achieve 100 mA/cm^2 is about 200 mV and the Tafel slope is 70 mV/dec.

7.2.2 Transition-Metal Phosphosulfides

There is also considerable interest in the development of transition-metal phosphosulfides (MPS), especially molybdenum phosphosulfides and cobalt phosphosulfides as highly active HER electrocatalysts. The P sites are postulated to be active for hydrogen adsorption and desorption. Hence, it is expected that transition-metal compounds containing both S and P will turn out to be excellent HER catalysts. Stain-induced d-band center shift due to the bigger atomic radium and electronic-effect-induced electronic structure tuning owing to the lower electronegativity of P than S play crucial roles in adjusting the catalysts' activity. DFT calculations indicate that replacement of S by P can decrease the electron occupation in antibonding e_g^*-orbitals due to the fewer valence electrons of P than those of S, thus resulting in the strengthening of chemical bonding between metal and ligands (S/P), and thus can enhance the chemical stability of the catalyst in the hydrogen evolution process. Jaramillo et al. first enhanced the HER activity and stability of MoP by introducing sulfur into the surface of MoP. They found that the synergistic effects between sulfur and phosphorus produced a high-surface-area electrode, which is more active than those based on either pure sulfide or phosphide. Therefore, the MoP with a phosphosulfide surface (MoP|S) exhibited one of the highest HER activities of any non-noble-metal electrocatalyst with a large exchange current density (0.57 mA/cm^2), high TOF value (0.75 s^{-1} at 150 mV), low overpotential (64 mV at the current density of 10 mA/cm^2), and low Tafel slope (50 mV/dec) (Fig. 7.6A,B). The incorporation of sulfur into the MoP surface mitigated surface oxidation of the phosphide, making the catalyst more active for the anodic-going sweep. Therefore, the MoP|S catalyst exhibited superb stability without a noticeable overpotential increase after 1000 potential cycles, whereas the MoP catalyst suffered a 10 mV increase of overpotential at the current density of 10 mA/cm^2. Tour et al. reported an improved HER activity of MoS$_2$ and MoP via the formation of molybdenum phosphosulfides (MoS$_{2(1-x)}$P$_x$) with different S/P ratios by annealing bulk particulate MoS$_2$ with different amounts of red phosphorus. The STEM image of MoS$_{0.94}$P$_{0.53}$ (Fig. 7.6C) becomes much more irregular and partially deformed as compared to that of MoS$_2$ (Fig. 7.6D), suggesting a strained basal plane after phosphorous incorporation. DFT calculations further indicate that the enhanced HER activity of the MoS$_{2(1-x)}$P$_x$ alloy is attributed to a much lower hydrogen adsorption free energy on the P-alloyed MoS$_2$ basal plane compared to that on pristine MoS$_2$ because electrons can be more easily filled when H binds on the P-alloyed MoS$_2$ surface (Fig. 7.6E,F).

Binary-Nonmetal TMCs for Hydrogen Evolution | 211

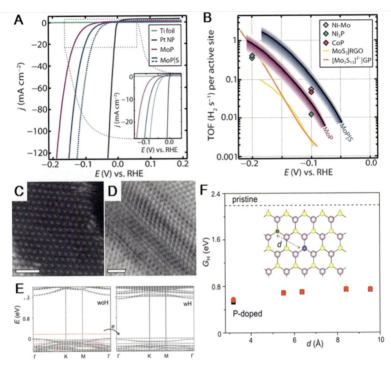

Figure 7.6 (A) The linear sweep voltammograms of MoP and MoP|S. The solid and dotted lines represent MoP|S samples with a loading of approximately 1 and 3 mg/cm², respectively. (B) TOF values of MoP and MoP|S together with TOFs of Ni–Mo, Ni$_2$P, CoP, MoS$_2$, and [Mo$_3$S$_{13}$]$^{2-}$ catalysts. (C) Higher resolution STEM image of the MoS$_2$ basal plane structure. (D) Higher resolution STEM image of MoS$_{0.94}$P$_{0.53}$ showing roughness and disorder on the surface. (E) Band structure of P-alloyed MoS$_2$ without (left) and with (right) H adsorption. (F) Free energy of H adsorption on the pristine MoS$_2$ surface (dashed line), and on the P-alloyed surface (dots) as a function of the in-plane distance between H and P.

Pyrite-type metallic cobalt disulfide (CoS$_2$) has emerged as an interesting material with high catalytic activity toward the HER. To further increase the HER activity and stability of the CoS$_2$ catalyst, phosphorus incorporation was performed to build highly active pyrite-type cobalt phosphosulfide (CoP|S) by Jin et al. Through P doping, the surface properties of CoS$_2$ could be optimized to more efficiently catalyze the HER with promoted charge transfer and facilitated proton adsorption. Different from CoS$_2$, which consists of Co^{2+} octahedra and S$_2^{2-}$ dumbbells, CoPS has a smaller lattice constant than CoS_2 and a composition of Co^{3+} octahedra and dumbbells with a homogeneous distribution of P^{2-} and S^- atoms. As the Co octahedra in CoPS contain P^{2-} ligands with higher electron-donating character than S^- ligands, after hydrogen adsorption at open P sites, the adsorption hydrogen free energy at the adjacent Co sites becomes spontaneous and almost thermoneutral owing to a spontaneous conversion between Co^{2+} and Co^{3+} (reduction of Co^{3+} sites to Co^{2+} on hydrogen adsorption at an adjacent P site and then oxidation of Co^{2+} back to Co^{3+} on subsequent hydrogen adsorption on Co sites). As a result, the CoPS electrodes achieved a catalytic current density of 10 mA/cm² at overpotentials as low as 48 mV (Fig. 7.7C). It is also observed that P substitution is critical to the chemical stability and catalytic durability of the material. In contrast to the parent CoS_2 catalyst, which suffered from a drastic decrease in current density by 70% in less than 30 min, the CoPS catalyst was able to sustain a current density of 10 mA/cm² for 20 h of continuous HER operation. Based on the projected DOS analysis, P substitution significantly influences the nature of the chemical bonding between Co and S/P and each Co atom is coordinated in an octahedral ligand field in the pyrite structure (Fig. 7.7D). Wang et al. inferred that when half of the S atoms are replaced by P, which has fewer valence electrons in the pyrite structure of CoS_2, the antibonding e_g^*-orbitals are depleted, which strengthens the chemical bonding between Co and ligands (S/P) and thus enhances the chemical stability of the catalyst in the hydrogen evolution process (Fig. 7.7E,F). The reason was confirmed to be that S and P can tune each other's electronic properties to produce an active catalyst phase.

In contrast to some half-metal TMCs (e.g., CoS_2), MPS have one less valence electron, making them semiconductors (e.g., CoPS). Integration of an electrocatalyst material with a conducting substrate typically leads to improvements in both performance and stability, since directly anchoring the electrocatalyst to a robust conducting support ensures a low resistance electrical transport pathway. Electronic coupling between the support and catalyst materials can synergistically boost intrinsic activity as well. Researchers have attempted to increase the electronic conductivity of metal phosphosulfides by directly growing MPS on conductive substrates, including carbon materials such as carbon fiber paper, graphene, Ni foam, etc. Ouyang et al. prepared P-doped CoS_2 nanosheet arrays

Binary-Nonmetal TMCs for Hydrogen Evolution | 213

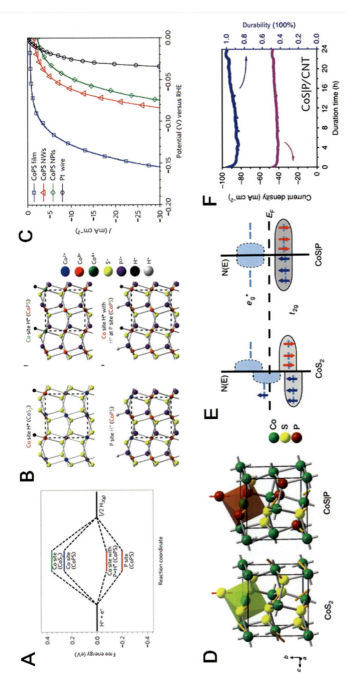

Figure 7.7 (A) Free-energy diagram and (B) schematic structural representations for hydrogen (H*) adsorption at the Co site on the (100) surface of CoS_2, and at the Co site, P site, and Co site after H* adsorption at the P site on the (100) surface of CoPS. (C) j–h curves after iR correction show the catalytic performance of the CoPS electrodes compared with that of Pt wire. (D) Structure illustration of pyrite-phase CoS_2 and CoS|P, each with a representative coordination polyhedron. (E) Conceptual energy level diagrams of the frontier molecular orbitals of pyrite-phase CoS_2 and CoS|P derived from the calculated electronic structures. (F) j–t curve recorded on the CoS|P/CNT electrode at a constant overpotential of 95 mV with iR correction.

on carbon fiber paper by phosphorization of CoS_2 nanosheets. Encouraged by the MoS_2/rGO system, which shows high activity for the HER, Sampath et al. composited layered ternary palladium phosphosulfides (PdPS) with reduced graphene oxide (rGO) for hydrogen evolution. Increase in the electronic conductivity of rGO–PdPS as compared to that of pure PdPS will help the electron-transfer kinetics for the HER, and thus lead to an ultralow Tafel slope of 46 mV/dec. Using a similar growth process to that of the rGO–PdPS system, the same group synthesized rGO–$FePS_3$, which contains both S and P as favorable hydrogen adsorption sites. The prepared rGO-few-layered $FePS_3$ composite exhibited decent HER performance with a Tafel slope of 45 mV/dec and an exchange current density of 1 mA/cm^2. The authors pointed out that the enhanced HER activity of the rGO-few-layered $FePS_3$ composite may have its origin in the improved conductivity due to the presence of rGO.

Like transition MPS, the formation of Se-doped or O-doped transition-metal phosphides modifies the nature of the d electronic structure of metals, giving rise to tunable HER catalytic properties. Jin et al. investigated the structure and HER activity of a sequence of pyrite-phase nickel phosphoselenide (NiPSe) nanomaterials. By adjusting the raw ratios of phosphorus and selenium, they synthesized a series of pyrite-phase nickel phosphoselenide materials—NiP_2, Se-doped NiP_2 ($NiP_{1.93}Se_{0.07}$), P-doped $NiSe_2$ ($NiP_{0.09}Se_{1.91}$), and $NiSe_2$—through a facile thermal conversion of $Ni(OH)_2$ nanoflakes. The results showed that Se-doped NiP_2 ($NiP_{0.09}Se_{1.91}$) showed the highest HER activity, which can achieve an electrocatalytic current density of 10 mA/cm^2 at an overpotential as low as 84 mV and a small Tafel slope of 41 mV per decade, providing another example of improving the HER catalytic activity by doping or alloying of nonmetal elements into the existing electrocatalysts. Qiao et al. developed a bifunctional catalyst electrode (Fe and O co-doped Co_2P–CoFePO) grown on nickel foam toward overall water splitting. It was believed that the atomic modulation between cations and anions plays an important role in optimizing the electrocatalytic activity, which greatly increased the active sites in the electrocatalyst. Very recently, Zhang et al. synthesized O-doped MoP and O-doped CoP nanosized particles, which are embedded in layered graphene (rGO) for the HER and OER, respectively (Fig. 7.8A). They found that the introduction of O atoms into transition-metal phosphides could not only enhance their intrinsic electrical conductivity, but also activate active sites via elongating the M–P bond, favoring the HER (Fig. 7.8B) and oxygen evolution reaction (Fig. 7.8C). First-principles calculations showed that O-doped TMPs could achieve a much higher DOS across the Fermi level, indicating enhanced intrinsic conductivity. Extended X-ray absorption fine structure spectroscopy (EXAFS) measurements show a slightly positive shift of Mo–P coordination for O-doped Mo–P compared to Mo–P (Fig. 7.8D) and of Co–P coordination for O-doped Co–P compared to Co–P (Fig. 7.8E), indicating the coordinative unsaturation or surface structural disorder in the presence of O atoms. As depicted in Fig. 7.8F, the ΔG_H of pure MoP is slightly far away from zero, while the ΔG_H of O-doped MoP is nearly zero at 3/4 monolayer H coverage for 1O–P–MoP or 2/4 monolayer H coverage for 2O–P–MoP systems. For the OER, the calculated adsorption energy of *OOH on CoO_x/CoP was 2.42 eV, while it decreased to 2.19 eV in the CoO_x/O-doped CoP system, implying a facile formation of *OOH after O incorporation (Fig. 7.8G).

7.2.3 Transition-Metal Carbonitrides

It has been subsequently demonstrated that carbides and nitrides of previous transition metals, which display Pt-like behavior in several catalytic reactions, are effective catalysts in hydrogen evolution. One of the primary interests in the applications of transition-metal carbides and nitrides was to use them as cheaper alternative catalysts to replace group VIII noble metals. It is well known that the electrocatalytic properties of transition-metal carbides and nitrides are related to the nature of the electronic structure of the materials. Theoretical band calculations have indicated that the electronic structure of these carbides and nitrides involves simultaneous contributions from metallic bonding related to the rearrangement of metal–metal bonds, covalent bonding due to the formation of covalent bonds between metal and nonmetal atoms, and ionic bonding characterized by the charge transfer between metal and nonmetal atoms. Among them the direction and amount of charge transfer and the modification effect on the metal d-band are two most important factors that influence their electronic structure.

216 *Application in Electrocatalytic Water Splitting*

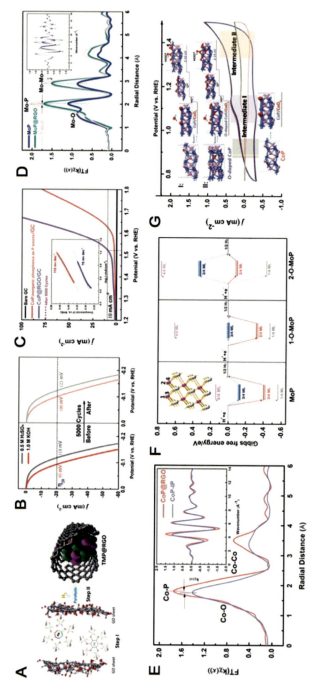

Figure 7.8 (A) Synthesis process of O-doped TMP@rGO by pyrolyzing phytic acid cross-linked complexes. (B) Free energy versus the reaction coordinate of the HER for various MoP models. (C) Polarization curves of MoP measured in both 0.5 M H_2SO_4 and 1 M KOH (left) and curves after 5000 potential sweeps at 20 mV/s (right). Mo (D) and Co (E) K-edge extended XAFS oscillation functions $k^3[\chi(k)]$ (inset) and their corresponding Fourier transform (FT). (F) Cyclic voltammetry (CV) responses of O-doped CoP (blue line) and pure CoP (red line) in N_2-saturated 1 M KOH. Inset shows the reaction Gibbs free energy for the OER on CoP. (G) Polarization curves in 1 M KOH with different forms of CoP catalysts, and curve (dashed line) after 5000 potential sweeps.

X-ray absorption analysis of carbon-supported Mo$_2$C revealed that the charge transfer from Mo to C downshifted the d-band center of Mo, thereby decreasing its hydrogen-binding energy. This effect, in turn, favors the electrochemical desorption of H$_{ads}$. The W–N bond has a more ionic character than metallic or carbidic W. A nitrogen-doping transition-metal carbide strategy was proposed by Zhao et al. to enhance the catalytic activity of tungsten carbide by further decreasing its hydrogen-binding energy. They prepared tungsten carbonitride nanoparticles with polydiaminopyridine (PDAP) as both carbon and nitrogen sources and Na$_2$WO$_4$ as the tungsten source (Fig. 7.9A). As iron is known to serve as a graphitization catalyst during pyrolysis, a small amount of iron was necessary to realize the synthesis of N-rich tungsten carbonitride (Fe–WCN). XPS shows that the binding energy of W downshifted with increase in the N atom content in the Fe–WCN materials (Fig. 7.9B). It was believed that the inclusion of nitrogen in WC further downshifted the d-band center of W, since the W atoms in WN are more electropositive than W atoms in WC. Therefore, the Fe–WCN catalysts displayed efficient HER activities in both acidic and alkaline media, with an HER overpotential of about 100 mV and a Tafel slope of about 47 mV/dec in acidic media. Inspired by the success of tungsten carbonitride, Zhao et al. turned their research to molybdenum carbonitrides (MoCN) as highly efficient HER catalysts. By using an in situ CO$_2$ emission strategy in which CO$_2$ gas bubbles were vigorously generated during the polymerization reaction, MoCN nanomaterials with high densities of catalytic active sites and effective surface areas were successfully synthesized (Fig. 7.9C). The obtained MoCN catalyst exhibited an HER onset potential of 50 mV and an overpotential of 140 mV to achieve the current density of 10 mA/cm^2. Hu et al. prepared N,P co-doped Mo$_2$C@C nanospheres by using phosphomolybdic acid as Mo and P sources and polypyrrole as N and C sources. The pomegranate-like Mo$_2$C@C nanospheres with a porous N-doped carbon shell as peel and N,P-doped Mo$_2$C nanocrystals as seeds exhibit extraordinary electrocatalytic activity for the HER in 1 M KOH electrolyte in terms of an extremely low overpotential of 47 mV at 10 mA/cm^2 and a Tafel slope of 68 mV/dec.

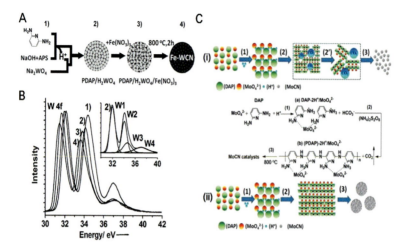

Figure 7.9 (A) Synthesis of Fe–WCN materials, where the gray dot is H$_2$WO$_4$, and the black dot is Fe(NO$_3$)$_3$. (B) W 4f spectra of Fe–WCN: (1) Fe–WCN-700, (2) Fe–WCN-800, (3) Fe–WCN-900, and (4) Fe–WCN-1000 catalysts prepared at various temperatures. The inset shows the deconvoluted XPS W 4f spectra of Fe–WCN-800, where W1 and W2 are N-bound W atoms, and W3 and W4 are O-bound W atoms. Panels (A and B). (C) Synthesis procedures of (i) PDAP-MoCN-CO$_2$ and (ii) PDAP-MoCN catalysts.

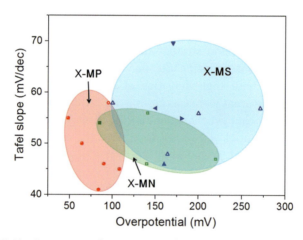

Figure 7.10 Summary and comparison of HER performance, including the overpotential at the current density of 10 mA/cm^2 and the Tafel slope of some typical binary-nonmetal TMC catalysts, such as foreign nonmetal doped transition-metal phosphides (X-MP), foreign nonmetal doped transition-metal nitrides (X-MN), and foreign nonmetal doped transition-metal sulfides (X-MS).

Sasaki et al. proposed a concept of hybridization of metal carbides and nitrides to enhance the HER catalysis. They reported graphene nanoplatelet-supported W_2C–WN nanocomposites via an in situ solid-state reaction. The X-ray absorption near edge structure (XANES) of $W_{0.5}$Ani/GnP showed further downshifting of the d-band center of tungsten in the presence of nitride compared with that of W_2C. The optimum ratio of nitride to carbide is 1 : 1, which gives a moderate M–H binding strength and thus enhances the HER activity. Incorporation of the W_2C–WN nanoparticles onto graphene nanoplatelets brings about a significant enhancement in the HER kinetics with a small overpotential of 120 mV at 10 mA/cm^2, and faster electron transport due to the remarkable reduction in the charge transfer resistance (12.7 Ω at η = 100 mV), and stable catalytic activity for producing hydrogen fuel over 300 h. The authors proposed that the presence of WN moderated the M–H binding strength on the material surface, inhibited the formation of tungsten oxide, and thus enhanced the HER activity. Xi et al. reported a facile nitrogenation/exfoliation process to prepare hybrid Ni–C–N nanosheets containing NiC and Ni_3N. The catalytic behavior of the Ni–C–N nanosheet is similar to that of the common Pt catalyst, and it can work well at all pH values for more than 70 h without an obvious current drop. The ultralow onset potential (34.7 mV) and overpotential (60.9 mV at a current density of 10 mA/cm^2) suggest that the Ni–C–N nanosheet catalyst is one of the most active non-Pt catalysts for the HER in acidic electrolytes ever reported at that time.

7.3 Some Basics of OER Catalysis

The basic criteria to evaluate OER catalysis will be introduced in this section. Overpotential (η) calculated by the difference between the applied potential (E) to reach a certain current density and the equilibrium potential (E_{eq}) (as shown in Eq. (7.8)) is the critical and most frequently used descriptor to evaluate the performance of an OER catalyst. E_{eq} is the half-reaction's thermodynamically determined reduction potential, and E is the potential at which the redox event is experimentally observed. Generally, η is usually given at the onset of OER and at the current density of 10 mA/cm^2, which corresponds to a 10% solar to hydrogen efficiency under 1 sun illumination. The existence of overpotential implies that the cell requires more energy than thermodynamically expected to drive a reaction. Therefore, a lower overpotential suggested a better OER activity.

$$\eta = E - E_{eq} \qquad (7.8)$$

Tafel slope (b), which describes the influence of potential/overpotential on steady-state current density, is another key descriptor to evaluate OER kinetics. The value of b could be calculated by Eq. (7.9), where R, T, and F are ideal gas constant, temperature, and Faradaic constant, respectively. α is the transfer coefficient, which is highly related to Tafel slope. It has been reported that if b = 120 mV/dec, the rate-determining step is dominated by the single-electron transfer step. If b = 60 mV/dec, it hints that the chemical reaction after one-electron transfer reactions is the rate-determining step. If b = 30 mV/dec, the rate-determining step is the third electron transfer step. Therefore, from the value of Tafel slope, we can roughly determine the rate-determining step of OER. Generally, small Tafel slope indicates fast reaction kinetics, and the rate-determining step is supposed to be at the ending part of the reaction. Therefore, catalysts with small Tafel slopes often show good catalytic activity for OER. However, it should be noted that the Tafel slope is often overestimated if geometric current density is used because of the fact that the geometric current density is usually smaller than the specific current density. Moreover, Tafel slope is not accurate to describe the performance of OER catalysts because of its oversimplified assumptions.

$$b = \frac{2.303RT}{\alpha F} \qquad (7.9)$$

Turnover frequencies (TOFs) refer to the turnover per unit time, representing the total number of moles transformed into the desired product by 1 mol of active site per time. Therefore, the number of TOFs determines the level of activity of the catalysts, which is given by Eq. (7.10), where j is the current density at a specified overpotential, A is the area of the electrode, and m is the number of moles of active materials deposited onto the electrodes. Moreover, it has been suggested that TOFs at different overpotentials could be different; therefore, the applied overpotential should be provided when presenting TOF.

$$\text{TOF} = \frac{jA}{4Fm} \tag{7.10}$$

Exchange current density (i_0) is defined as the current density at $\eta = 0$ (j_0) divides surface area (A); the magnitude of i_0 reflects the intrinsic charge transfer between reactant and catalyst (Eq. (7.11)). A higher i_0 hints at better catalytic performance. i_0 can be described by Eq. (7.12), in which $k°$ is the rate constant, and ρ and ω are the reaction orders of Red and Ox, respectively.

$$i_0 = \frac{j_0}{A} \tag{7.11}$$

$$i_0 = k°[\text{Red}]^\rho [\text{Ox}]^\omega \tag{7.12}$$

Different from exchange current density, geometric current density (jg) is given by the current density normalized by geometric surface area at a certain overpotential. jg has practical meaning in developing water-splitting devices; however, it usually overestimates the electrochemical performance of a catalyst due to the larger actual surface area than the geometric surface area. Figure 7.11 shows the Tafel slopes of Co_3O_4 and IrO_2 by using the current density calculated by Brunauer–Emmett–Teller (BET), electrochemical (EC), and disk surface area, respectively. From Fig. 7.11, we can also see that the Tafel slope of Co_3O_4 is smaller than that of IrO_2 with respect to the geometric surface area. One may rush to conclude that the OER performance of Co_3O_4 is better than IrO_2. However, if the BET or EC surface area is used, the performance of Co_3O_4 is poorer than that of IrO_2, suggesting that the applied active surface area is quite important to determine the performance of a catalyst. Generally, the more accurate the surface area used, the more accurate the current density that is obtained, leading to a more precise evaluation of the catalyst.

Understanding the mechanism of OER has a fundamental importance in designing new OER catalysts; therefore, let us first discuss the general mechanisms of OERs before presenting the detailed reactivity about LDH. The catalytic cycle is shown in Fig. 7.12; association mechanism and oxy–oxy coupling mechanism are generally proposed. There are four elementary steps for the association mechanism (Eqs. (7.13)–(7.16)): association of hydroxide anions to form absorbed OH* accompanied by losing one electron, generation of reactive oxy intermediate O* from OH* with loss of one electron and generation of one molecule of water, nucleophilic attack of absorbed oxy O* by the hydroxide anion with release of one electron to form O–O bond giving OOH*, and formation of one molecule of oxygen with release of an electron and one molecule of water to regenerate the catalyst and complete the catalytic cycle. For the oxy–oxy coupling mechanism (Eqs. (7.13), (7.14), and (7.17)), one molecule of oxygen will be generated accompanied by the regeneration of catalyst after generating the oxy intermediate O*. For the association mechanism, the formation of OOH* is generally regarded as the rate-determining step due to the large energy barrier according to density functional theory (DFT) calculations. For the oxy–oxy coupling mechanism (given by Eqs. (7.13), (7.14), and (7.17)), the coupling between two oxy is supposed to withstand a very high kinetic barrier and thus is the rate-determining step.

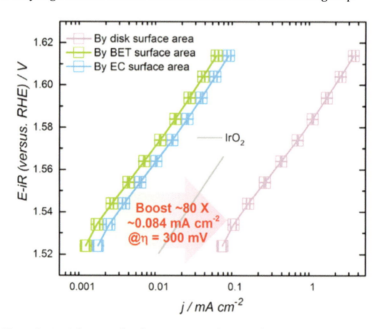

Figure 7.11 Influence of surface area on evaluation of a catalyst. Tafel plots of OER on IrO_2 and Co_3O_4 made at 300°C, in which the OER currents are normalized by the disk surface area, BET surface area, and EC surface area, separately. (Wang et al., 2018).

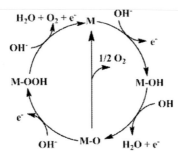

Figure 7.12 Catalytic cycle for the OER on transition-metal-based catalysts in alkaline conditions.

In the aforementioned processes, the formation of OOH* involves oxidation of oxygen from O* to OOH*, which is usually regarded as the rate-determining step. Therefore, LDH with high oxidation ability would facilitate the formation of OOH*. In addition, OER involves formation and cleavage of metal–oxygen bonds; in principle, catalysts with superior OER activity should possess a suitable oxygen-bonding strength, neither too strong nor too weak. As oxidation ability and oxygen-binding energy of LDH vary with the change in transition metals, they have a critical influence on the OER activity of LDH. Therefore, in this perspective, we will focus on the effects of chemical composition on the oxygen-binding energy and oxidation ability of LDH. Unary, binary, and ternary transition-metal-based LDHs toward OER will be discussed.

$$M + OH^- \rightarrow M-OH + e^- \qquad (7.13)$$

$$M-OH + OH^- \rightarrow M-OH + H_2O + e^- \qquad (7.14)$$

$$M-O + OH^- \rightarrow M-OOH + e^- \qquad (7.15)$$

$$M-OOH + OH^- \rightarrow M + H_2O + O_2 + e^- \qquad (7.16)$$

$$M-O + M-O \rightarrow M + O_2 \qquad (7.17)$$

7.4 Transition-Metal-Based Layered Double Hydroxides for Oxygen Evolution

7.4.1 Unary Metal-Based Layered Double Hydroxides

Unary metal-based layered double hydroxides (LDHs) exhibited limited OER activity, but they provide an ideal platform for us to understand the intrinsic OER activity of LDHs due to their structural simplicity (Wang et al., 2018; Long et al., 2016). In this part, we will first introduce Ni-based LDH, followed by Fe-based LDH, Co-based LDH, and the recently reported V-based LDH. As transition-metal hydroxide and transition-metal-oxy/hydroxide can interconvert via Bode's diagram, we also treated transition-metal oxyhydroxide as LDH for simplicity.

7.4.1.1 VIII group single transition-metal hydroxides/oxyhydroxides

Ni-based compounds are the most widely used OER catalysts; actually, NiO_x was employed for OER early in the 1980s. However, it did not arouse research interests until 2012 when Boettcher, in situ, generated nickel layered hydroxide/oxyhydroxide from NiO_x through an electrochemical conditioning process (Fig. 7.13). The as-in situ-generated Ni hydroxide/oxyhydroxide exhibited an outstanding OER performance, with a low overpotential of 297 mV at 1 mA/cm², an extremely small Tafel slope of 29 mV/dec, and a considerably large TOF of 0.17 s⁻¹ at η = 300 mV in 1 mol/L (M) KOH, better than that of the state-of-the-art catalyst of IrO_x catalysts (η = 378 mV at 1 mA/cm², b = 49 mV/dec, TOF = 0.0089 s⁻¹ at η = 300 mV in 1 M KOH).

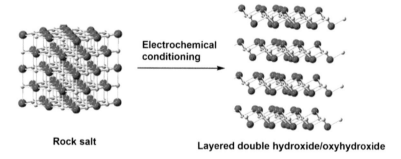

Figure 7.13 Proposed in situ transformation from the thermally prepared oxides to the layered hydroxide/oxyhydroxide structure from NiO_x.

NiFe LDH is most effective for OER, but NiOOH and $Ni(OH)_2$ are not so effective toward OER. Therefore, studying the OER activity of Fe-based LDH is important to understand the superior activity of NiFe LDH and has aroused much attention. Friebel and Bell studied the intrinsic OER activity of γ-FeOOH and found that the overpotential at 10 mA/cm^2 of γ-FeOOH is 550 mV in 0.1 M KOH, which is smaller than that of Fe-free γ-NiOOH (η = 660 mV at 10 mA/cm^2 in 0.1 M KOH). However, it is much higher than that of (Ni,Fe)OOH (η = 360 mV at 10 mA/cm^2 in 0.1 M KOH). Moreover, the calculations also indicated that the overpotential of γ-FeOOH is 520 mV, in good agreement with experiments. Boettcher also studied the OER activity of FeOOH and suggested that FeOOH had high OER activity, but was limited by its poor conductivity, which has a measurable conductivity of 2.2 × 10^{-2} mS/cm only when the overpotential is larger than 400 mV.

Similar to Ni and Fe, the remaining first-row group VIII transition metal, cobalt, can also form a LDH structure and, of course, has received much interest. Wang compared the OER activities of α-Co(OH)$_2$, β-Co(OH)$_2$, and β-CoOOH and found that α-Co(OH)$_2$ will transform to γ-CoOOH before the OER, and the resulting γ-CoOOH inherits a large basal distance of α-Co(OH)$_2$. It has an overpotential of 400 mV at 10 mA/cm^2 in 0.1 M KOH; interestingly, a Tafel slope of 44 mV/dec when η is smaller than 350 mV, and 130 mV/dec when η is larger than 350 mV. Moreover, α-Co(OH)$_2$ is more active than β-Co(OH)$_2$, which might be due to the large interlayer space in α-Co(OH)$_2$.

Materials with ultrathin structures usually have high specific and exposed surface area and are abundant in vacancies, leading to a higher number of active sites and thus higher activities. Pan and Wei synthesized an atomically thin γ-CoOOH, with a thickness of only 1.4 nm. Expectedly, the as-prepared γ-CoOOH has very high mass activities and abundant active sites, thus leading to a sharp increase in OER activity with η = 300 mV at 10 mA/cm^2 and a Tafel slope of 38 mV/dec in 1 M KOH. Interestingly, the as-prepared γ-CoOOH is half-metallic in contrast to its bulk, which was proposed to be related to the presence of dangling bonds in the CoO$_{6-x}$ octahedron as supported by DFT calculations. Kang and Yao also prepared Co-based LDH with atomic thickness for OER. Owing to its ultrathin structure, the OER activity of the Co-based LDH can have an overpotential of 340 mV at 10 mA/cm^2, a Tafel slope of 56 mV/dec, and a TOF of 0.801 s^{-1} at η = 350 mV in 1 M KOH. Besides ultrathin structures, a larger interlayer spacing of LDH can also lead to a higher number of active sites. Sun and Chen reported benzoate anion interacted with CoOOH with an interlayer spacing as large as 14.72 Å, which allows the easy permeation of water and hydroxide, resulting in a higher number of active sites. And the overpotential at 50 mA/cm² is only 291 mV in 1 M KOH.

7.4.1.2 V-hydroxides/oxyhydroxides

In addition to the intensively studied VIII transition-metal-based LDH, V and Mn compounds have also been investigated. In 2012, Markovic studied the trends for OER on 3d transition-metal hydr(oxy)-oxide catalysts ($M^{2+\delta}O_\delta(OH)_{2-\delta}$/Pt(111)) and discovered that the reactivity toward OER is in the order Mn < Fe < Co < Ni, which is governed by the OH-$M^{2+\delta}$ bond strength (Ni < Co < Fe < Mn). According to the Sabatier principle, too weak or too strong M–OH bonds would retard the OER reactivity. Copper and zinc have too many d electrons in the d-orbitals, which cause strong repulsions between the d electrons and 2p electrons of oxygen. Expectedly, $Cu(OH)_2$ and $Zn(OH)_2$ exhibited poor OER activities. Early transition metals, such as titanium, have few d electrons in the d-orbitals, which lead to very strong OH–M bonds, and are supposed to be ineffective toward OER.

Despite that the early transition-metal hydroxide has a strong M–OH bond, which is proposed to be unfavorable for OER, VOOH hollow nanospheres, structurally resembling lepidocrocite γ-FeOOH, have been employed as an efficient OER catalyst by Liang and Wang, exhibiting an overpotential of 270 mV for OER at 10 mA/cm² and a Tafel slope of 68 mV/dec in 1 M KOH. It has been well established that V is an early transition metal, favoring its high oxidation states, +5 and +4, whereas V has an oxidation state of +3 in VOOH. And the reactivity of VOOH does not decrease even after 5000 cycles. Moreover, the prepared VOOH can be used as HER catalyst, with an overpotential of 164 mV at 10 mA/cm² and a Tafel slope of 104 mV/dec. The advanced water-splitting performance of VOOH in this work was contributed to the large surface area from the hollow sphere morphology.

7.4.2 Binary Metal-Based LDH

Unary-transition-metal-based LDHs are limited by their low intrinsic OER activity or conductivity. Fortunately, by doping the second metal ions into the unary-TM LDH, the as-formed binary LDHs such as NiFe LDH, NiCo LDH, and CoFe LDH showed a much higher OER performance, which will be summarized and discussed in this section.

7.4.2.1 NiFe LDH

LDH suffers from low conductivity, which is one of the major challenges needed to be overcome for an advanced OER performance. Therefore, various methods have been applied to tackle this problem. In 2013, Dai deposited NiFe LDH on carbon nanotubes (CNTs) to obtain NiFe-CNT LDH based on the fact that the end tips of CNT have many carboxylic groups that can coordinate to metal centers (Fig. 7.14a). The as-prepared NiFe-CNT LDH exhibited a Tafel slope of 31 mV/dec at 1 M KOH and has an overpotential of 290 mV at 5 mA/cm^2. The TOF is 0.56 s^{-1} at an η = 300 mV in 1 M KOH. Moreover, the NiFe-CNT LDH exhibited a better stability than Ir/C at a constant current density of 5 mA/cm^2. In 2014, Yang reported the graphene oxide (GO) intercalated NiFe LDH by substitution of anions (CO$_3^{2-}$ or Cl$^-$) with GO (Fig. 7.14b). The as-prepared NiFe-GO LDH has an overpotential as low as 210 mV at 10 mA/cm^2 and a Tafel slope of only 40 mV/dec in 1 M KOH. Moreover, the TOF can reach 0.38 s^{-1} at η = 300 mV. Since the conductivity of GO can be enhanced by reduction, we further reduce NiFe-GO LDH by hydrazine to obtain reduced GO intercalated NiFe-reduced graphene oxide (rGO) LDH; the overpotential and Tafel slope can be further lowered to 195 mV at 10 mA/cm^2 and 39 mV/dec, respectively. Moreover, the TOF can reach 0.98 s^{-1} at η = 300 mV. The higher conductivity of FeNi-rGO LDH is evidenced by alternating current impedance spectra. It should be noted that the basal spacing of FeNi-GO LDH is 1.1 nm, which is much larger than the basal spacing of FeNi-CO$_3$ LDH (0.75 nm), suggesting that GO has successfully intercalated into the NiFe LDH. And the enlarged basal distance of NiFe LDH allows efficient association of reactants and dissociation of products, and that is why a sharp increase in TOF is observed. Later, NiFe LDH-GO and NiFe LDH-rGO have also been reported through solvothermal and chemical reduction methods by Zhan and Hou.

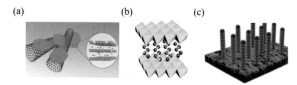

Figure 7.14 Schematic layout of the structure of NiFe-CNT LDH (a), NiFe-GO LDH (b), and NiFe LDH on conducting Ni_3S_2 nanorods (c).

Besides combination with conductive carbon materials, other conducting materials were also applied to reduce the charge transfer resistance of LDHs. Wang and Qiu prepared hierarchical NiFe LDH/Ti_3C_2-MXene for OER. Owing to the stabilization by Ti_3C_2-MXene, better conductivity, and electronic interaction between Ti_3C_2-MXene and NiFe LDH, the as-prepared hierarchical NiFe LDH/Ti_3C_2-MXene exhibited a Tafel slope of 43 mV/dec and overpotential of ~300 mV at 10 mA/cm^2 in 1 M KOH. The TOF at η = 300 mV is 0.26 s^{-1}. Zhang and Huang grew NiFe LDH on conducting Ni_3S_2 nanorods, as shown in Fig. 7.14c. As Ni_3S_2 nanorods are much more electron rich than NiFe LDH, electrons will flow to NiFe LDH, leading to the partial reduction of NiFe LDH. To maintain the charge neutrality, oxygen vacancies are generated in NiFe LDH as evidenced by X-ray photoelectron spectroscopy (XPS). As a result, the intrinsic activity of NiFe LDH was enhanced. Moreover, the number of active sites also increased after being loaded on the Ni_3S_2 nanorods. The overpotential at 10 mA/cm^2 is as low as 190 mV, and the Tafel slope is also only 38 mV/dec in 1 M KOH. Wang reported NiFe LDH@Au hybridized nanoarrays on nickel foam. The as-prepared NiFe LDH@Au/Ni foams exhibited overpotentials of only 221, 235, and 270 mV at 50, 100, and 500 mA/cm^2, respectively, in 1 M KOH. Moreover, the Tafel slope decreased to 48.4 mV/dec, in comparison to NiFe LHD/Ni foam with a Tafel slope of 71.1 mV/dec. Huang reported single-crystalline NiFe LDH array on an Ni foam, and the as-prepared NiFe LDH array showed an excellent OER activity. The overpotentials at 10, 50, and 100 mA/cm^2 are only 210, 240, and 260 mV in 1 M KOH, which is smaller than the coated NiFe LDH film. Moreover, the Tafel slope is 31 mV/dec in the overpotential region of 240–260 mV, suggesting a faster OER kinetics and large current density. Xie, Zheng, and Sun grew an amorphous NiFe-borate layer on an NiFe LDH surface and found that the OER performance was greatly boosted owing to the higher surface roughness and increased number of active sites. NiFe LDH is effective toward OER but suffers from lower conductivity. It has been reported that substitution of the very electronegative oxygen to other less electronegative elements, such as sulfur, selenium, phosphorus, and nitrogen, can push up the valence bond of NiFe LDH, thus leading to a higher conductivity and a better OER activity.

7.4.2.2 Other Ni-based binary metal LDHs

Besides NiFe LDHs, other Ni-based bimetal LDHs, including NiCo, NiMn, NiCr, NiTi, NiV, NiGa, and NiAl LDHs, also have been well studied, which will be discussed in this section.

Qian and Li reported NiCo LDH nanosheet arrays on Ni foam for overall water splitting. The as-prepared NiCo LDH has an overpotential of 271 mV at 10 mA/cm^2 and a Tafel slope of 72 mV/dec in 1 M KOH. And it should be noted that it was used as a bifunctional catalyst for overall water splitting for the first time, which showed an overpotential of 162 mV for HER at a current density of 10 mA/cm^2. Similarly, Jiang and Ai reported NiCo LDH nanosheets for OER that has an overpotential of 290 mV at 10 mA/cm^2, with the Tafel slope being 113 mV/dec in 1 M KOH. And XPS characterizations suggested Co^{3+} and Co^{2+} coexist in the prepared NiCo LDH. NiCo LDH nanosheet has also been reported by Huang, which shows an overpotential of 282 mV at 10 mA/cm^2 in 1 M KOH, and the Tafel slope is 42.6 mV/dec. By post-exfoliation treatment, Hu reported a single-layered NiCo LDH, which showed a much increased OER activity than the bulk counterpart. The overpotential at 10 mA/cm^2 and Tafel slope decreased from ~390 mV and 65 mV/dec to ~334 mV and 41 mV/dec^1, respectively, whereas the TOF increased from 0.025 to 0.01 s^{-1}.

The early or middle transition-metal-Ni LDHs have also been studied, but admittedly, they are relatively less studied compared with NiFe LDH and NiCo LDH because the formed M–OH and M–O bonds are too strong, which is unfavorable for OER. For example, Sun reported NiMn LDH for OER with an overpotential of 640 mV at 20 mA/cm^2 in 1 M KOH; however, it was less effective than NiFe LDH (401 mV at 20 mA/cm^2 in 1 M KOH). NiMn LDH nanosheet has also been synthesized by Huang for overall water splitting. However, in this work, the NiMn LDH with an overpotential of 312 mV at 10 mA/cm^2 is less effective than NiFe LDH with an overpotential of 220 mV at 10 mA/cm². NiTi LDH, which acts as a precursor to prepare NiO-TiO$_2$ ultrafine nanosheet, is ineffective for OER. The Tafel slope is as high as 290 mV/dec, and the TOF at an overpotential of 500 mV is only 0.009 s^{-1}. Surprisingly, monolayer NiV LDH, which shows a comparable OER activity with NiFe LDH, has been reported by Sun. V exists in V^{3+}, V^{4+}, and V^{5+}, as evidenced by XPS shown in Fig. 7.15a. The presence of V^{4+} and V^{5+} is attributed to the oxidation of V^{3+} during the synthesis. And the as-prepared monolayer NiV LDH (Ni$_{0.75}$V$_{0.25}$-LDH) has an overpotential of 250 mV, a Tafel slope of 50 mV/dec, and a TOF of 0.054 ± 0.003 s^{-1} in 1 M KOH, whereas the related monolayer NiFe LDH (Ni$_{0.75}$Fe$_{0.25}$-LDH) has an overpotential of 300 mV, Tafel slope of 0.064 mV/dec, and a TOF of 0.021 ± 0.003 s^{-1}. Electrochemical impedance spectroscopies (EISs) indicated that Ni$_{0.75}$V$_{0.25}$-LDH has a lower charge transfer resistance, thus bearing with a high conductivity. Moreover, the as-prepared NiV LDH exhibited considerable stability, as shown in Fig. 7.15b. DFT calculations have been performed on the OER mechanism catalyzed by Ni$_{0.75}$V$_{0.25}$-LDH. Here, V is supposed to be the active site, where H$_2$O*, OH*, O*, OOH*, and OO* bind on V, and the rate-determining step is the formation of OOH* from O*, which has an overpotential of 620 mV.

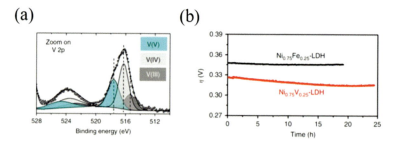

Figure 7.15 Zoom on the V 2p core-level XPS measurements of NiV LDH power deposited on a fluorine-doped tin oxide (a); long-term stability test of NiV LDH and NiFe LDH (b).

In addition, some main-group element-Ni LDHs are also reported; as expected, their OER activity is very low. To synthesize porous β-Ni(OH)$_2$, Wang and Jin first prepared NiGa LDH in which the Ga^{3+} will be removed through base etching. And it was found that the prepared NiGa LDH exhibited a little better OER activity than the β-Ni(OH)$_2$ nanosheet, but it was much poorer than the porous β-Ni(OH)$_2$ prepared by etching NiGa LDH. Similarly, the NiAl LDHs are also prepared as a precursor to synthesize the porous LDH. For example, Zhang and Xie employed NiAl LDH as a precursor to prepare β-Ni(OH)$_2$ ultrathin nanomesh. The prepared β-Ni(OH)$_2$ ultrathin nanomesh has an excellent OER activity, whereas the NiAl LDH only shares a similar OER activity with the β-Ni(OH)$_2$ nanosheet.

7.4.2.3 Co-based binary metal LDH

Cobalt, with one less d electron than nickel, also received much attention. Li and Ge reported CoFe LDH fabricated by coprecipitation, in which the ratio of Co to Fe can be tuned from 0.5 to 7.4. The Co$_2$Fe LDH gave the best OER activity with an overpotential of 290 mV at 10 mA/cm^2 and a Tafel slope of 83 mV/dec in 1 M KOH. Again, ultrathin LDHs with abundant defects are supposed to have higher OER activities. Wang prepared ultrathin CoFe LDHs through Ar plasma etching. The Ar plasma etched has a thickness of 0.6 nm, compared to the bulk with a thickness of 20.6 nm. The ultrathin CoFe LDHs are abundant in Co, Fe, and O vacancies, as evidenced by the decreased coordination numbers of Co, Fe, and O. The ultrathin CoFe LDH-Ar has a TOF of 4.78 s^{-1}, whereas the bulk CoFe LDH has a TOF of 1.12 s^{-1}. And the Tafel slope and overpotential at 10 mA/cm^2 of CoFe LDH-Ar are 37.85 mV/dec and 266 mV, respectively, which are much lower than those in bulk CoFe LDH (57.05 mV/dec and 321 mV at 10 mA/cm^2). It should be noted that ultrathin CoFe LDH-Ar has a much smaller charge transfer resistance. Xiong and Sun reported an ultrathin CoFe-borate-layer-coated CoFe LDH nanosheet array supported on Ti mesh. The as-prepared catalysts showed good OER activity at near-neutral condition (0.1 M K$_2$B$_4$O$_7$ solution, pH = 9.2), with an overpotential of 418 mV at 10 mA/cm^2.

Boettcher studied the roles of Co and Fe in CoFe LDH. Compared with CoOOH (rigorously Fe-free) films that exhibited a TOF of 0.007 ± 0.001 s^{-1} and FeOOH films that showed a TOF of 0.016 ± 0.003 s^{-1}, CoFeOOH with x between 0.4 and 0.6 has a high TOF up to 0.61 ± 0.10 s^{-1}. Moreover, the Tafel slope decreased to 26–39 when x was in the range from 0.33 to 0.79, from 62 mV/dec for CoOOH to 45 mV/dec for FeOOH (Fig. 7.16). Moreover, it has been found that the Co$^{2+/3+}$ wave shifts anodically with the increase in Fe amount. Considering the low conductivity and unstable character of FeOOH under the OER process in alkaline condition, CoFe LDH with the incorporation of Co had a higher conductivity, hence showing a higher OER activity.

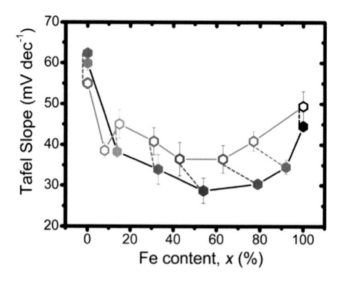

Figure 7.16 Tafel slopes from the second cyclic voltammetry cycle (10 mV/s) taken before (solid circle) and after (open circle) a 2 h polarization at η = 350 mV. Dotted lines link the pre- and post-polarization values.

Hu reported ultrathin CoMn LDH (3.6 nm) for OER; the as-prepared CoMn LDH has an overpotential of 325 mV at 10 mA/cm^2 and Tafel slope of 43 mV/dec in 1 M KOH, better than the sum of Co(OH)$_2$ and Mn$_2$O$_3$. Interestingly, after anodic conditioning, the overpotential can be further reduced to 293 mV, which is proposed to be related to the accumulation of Co(IV) species in the amorphous layers. Recently, Cheng and Liu reported strongly electrophilic Mn^{4+}-doped CoOOH nanosheet (i.e., CoMn LDH) for OER. Theoretical calculations indicated that incorporation of Mn^{4+} leads to higher occupancy at the Fermi level (mainly conduction band), thus facilitating electron transfer in the CoMn LDH. Moreover, it was found that incorporation of Mn^{4+} enhanced the binding of OH$^-$ to Co by 0.7 eV. Owing to the increased conductivity and stronger OH$^-$ binding energy, CoMn LDH exhibited a higher OER activity with η = 255 mV at 10 mA/cm^2 and Tafel slope of 38 mV/dec in 1 M KOH.

Considering that Co^{2+} is the OER active site and Cr^{3+}-based oxides always exhibit good conductivity, Huang synthesized CoCr LDH for OER aiming for a good conductivity and OER activity. Indeed, the conductance of CoCr LDH is 4.5 times and 21.4 times higher than that of CoOOH and $Co(OH)_2$, respectively. Different atomic ratios of Co and Cr were studied, and it was found that the Co_2Cr LDH gave the best OER activity with an overpotential of 240 mV at 10 mA/cm^2 and a Tafel slope of 81.0 mV/dec, which are the most active among the best of Co-based candidates. And the high activity was proposed to be contributed by the modified electronic structure, improved surface areas, and better conductivity introduced by Cr^{3+}. Asefa prepared ZnCo LDH for water and alcohol oxidation in which the ratio of Co^{3+} to Co^{2+} was 1 and the ratio of Co to Zn was also 1. The overpotential of ZnCo LDH is 0.34 V in 0.1 M KOH solution, which is lower than that of Co_3O_4 and CoOOH. The TOF at an η of 700 mV can reach up to 0.88 s^{-1}. It should be noted that ZnCo LDH has a much smaller Faradaic impedance than Co_3O_4. Although it was suggested that Zn is inactive for OER and Co is the active center, the Zn^{2+} was proposed to facilitate the formation of highly oxidized Co ions in ZnCo LDH, similar to the role of Ca^{2+} in $[Mn_3CaO_4]^{6+}$ catalyst. Later on, Xiang and Yan first mixed $ZnSO_4$ and $CoSO_4$ and then added some H_2O_2 to oxidize Co^{2+} to Co^{3+}, followed by electrodepositing it to a three-electrode configuration to prepare ZnCo LDH. The prepared ZnCo LDH has an overpotential of 427 mV at 2 mA/cm^2 in 1 M KOH (510 mV at 10 mA/cm^2), and the Tafel slope is 83 mV/dec. The TOF at η = 700 mV is 3.56 s^{-1}, much higher than that of ZnCo LDH prepared by the coprecipitation method. CoAl LDH has been reported as a precursor to prepare Al-doped CoP nanoarrays for overall water splitting.

7.4.3 Ternary Metal-Based LDH

Although binary metal-based LDHs exhibited an improved OER activity than unary metal-based LDH, they often suffer from poor conductivity. One may think incorporation of a third metal could involve new states in the forbidden band of binary metal-based LDH, thus leading to a higher conductivity. Moreover, introduction of another metal might increase the number of active sites. By the way, there are some examples of Ru- and Ir-doped NiFe LDHs, but with the focus on the HER in alkaline condition, which will not be discussed here.

In 2014, Yang reported ultrathin NiCoFe LDH for OER. The prepared Ni_8Co_2Fe LDH has a low overpotential of 210 mV at 10 mA/cm^2, a Tafel slope of 42 mV/dec, and TOF of 0.7 s^{-1} in 1 M KOH at an overpotential of 300 mV, outperforming the $Ni_{10}Fe$ LDH with η = 210 mV at 10 mA/cm^2, Tafel slope of 55 mV/dec, and TOF of 0.53 at η = 300 mV. The specific surface area of Ni_8Co_2Fe LDH is 80.44 m^2/g, much larger than that of $Ni_{10}Fe$ LDH (46.05 m^2/g), suggesting that Co incorporation might lead to more exposed active sites. In addition, charge transfer resistance also decreased a lot upon incorporation of Co. Inspired by the fact that Ni^{2+}/Co^{2+} confinement in the interlayer space of birnessite enhanced the OER activity, Yan and Strongin prepared the cobalt-intercalated NiFe LDH aiming for a better OER activity. As the Co–O band length is found to be between CoOOH and Co(OH)$_2$ based on extended X-ray absorption fine structure fit results, Co^{2+} and Co^{3+} coexist. The performed DFT calculations indicated that substitution of Ni^{2+} by Co^{2+} can lower the overpotential of OER (780 mV) by decreasing the Gibbs free energy differences between O* and OOH*, whereas substitution of Fe^{3+} by Co^{3+} leads to even lower overpotential (680 mV) of OER. The higher OER activity is thought to be the modified binding strengths of O* and OOH* due to the hybridization of 3d-orbitals of Co and 2p-orbitals of O at the valence band maximum. NiCoFe LDH has also been reported by Reguera with an overpotential of 250 mV at 10 mA/cm^2. In this contribution, Co^{3+} in low-spin configuration was thought to act as a shield of iron effect over nickel, which modulates the necessary potential for nickel oxidation.

Further, Sun reported that ternary NiFeMn LDH for OER with an overpotential of 289 mV at 20 mA/cm^2 in 1 M KOH is more efficient than NiMn LDH, which has an overpotential of 640 mV at 20 mA/cm^2. Moreover, the ternary NiFeMn LDH is also more effective than NiFe LDH (401 mV at 20 mA/cm^2). The higher OER activity of NiFeMn LDH compared to NiFe LDH was attributed to the fact that Mn^{4+} modifies the electronic structures of NiFe LDH, hence leading to a higher conductivity, as evidenced by DFT calculations and sheet resistance.

Liu and Sun reported NiFeV LDH with an overpotential of only 195 mV at 20 mA/cm² and a Tafel slope of 42 mV/dec in 1 M KOH solution, much better than NiFe LDH (η = 249 mV at 20 mA/cm², b = 49 mV/dec in 1 M KOH) and NiV LDH (η = 330 mV at 20 mA/cm², b = 72 mV/dec in 1 M KOH). The higher OER activity of NiFeV LDH was ascribed to the increased conductivity modified by V doping, as evidenced by DFT calculations and EIS. In addition, the ECSA of NiFeV is also larger than NiFe LDH, suggesting a higher number of active sites of NiFeV LDH.

To prepare NiFe LDH with atomic-scale defects, Kuang and Sun first prepared NiFeZn and NiFeAl LDH precursors; then, the Al and Zn were partially removed by alkaline etching to obtain defects containing NiFeZn and NiFeAl LDH, denoted D-NiFeZn LDH and D-NiFeAl LDH, respectively. D-NiFeAl LDH has a lower OER activity than NiFe LDH, but higher activity than NiFeAl LDH. Interestingly, D-NiFeZn LDH shows a much higher OER activity than NiFe LDH and NiFeAl LDH, with an overpotential of only 200 mV at a current density of 20 mA/cm² in 1 M KOH, and the resulting D-NiFeZn LDHs are abundant in the Ni-O-Fe unit, which is regarded as the active site. DFT calculations suggested that formation of O* from OH* is the rate-determining step, and the D-NiFeZn LDH has a lower overpotential than D-NiFeAl LDH. Interestingly, Rezvani and Habibi reported that ternary NiFeZn has a superior OER activity than binary NiFe LDH. Moreover, the OER can take place at neutral condition, and the Tafel slope can be reduced to as low as 16 mV/dec, whereas the NiFe LDH has a Tafel slope of 29 mV/dec. However, the current density is only 5.41 mA/cm² at an overpotential of 300 mV. The better OER performance of ternary NiFeZn was attributed to its higher conductivity modified by Zn^{2+}.

7.5 Mechanistic Studies of OER

Although transition-metal-based OER catalysts have been studied experimentally, the theoretical investigation of their structure–composition–performance relationship is far from satisfactory because of the inaccurate models. Actually, only the structure of β-Ni(OH)₂ has been accurately determined, which belongs to the P3⁻m1 (brucite) space group, and the lattice parameters are $a = b = 3.12$ Å and $c = 4.66$ Å. Carter performed DFT + U calculations and found that the pure β-NiOOH has a proton-staggered structure and is antiferromagnetic (Fig. 7.17).

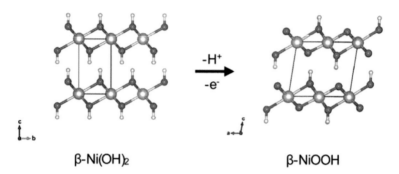

β-Ni(OH)₂   β-NiOOH

Figure 7.17 Experimental structure of β-Ni(OH)₂ and calculated structure of β-NiOOH.

Studying the surfaces of catalysts has fundamental importance because most of the OERs take place at the surfaces. Different surfaces of the catalyst could lead to a different reactivity. As β-NiOOH shares a similar backbone with LDH and can be modified to be a more efficient OER catalyst, Carter performed theoretical calculations on the stability and chemistry of β-NiOOH at the vacuum and aqueous conditions. The calculations suggested that the surface stability in the vacuum follows the order (0001) > {01$\bar{1}$N} ≫ {01$\bar{2}$N}, N = 0, 1 due to the presence of dangling bonds at {01$\bar{1}$N} and {01$\bar{2}$N} surfaces. However, the order of stability becomes (0001) > {01$\bar{1}$N} ≈ {01$\bar{2}$N} in aqueous condition, owing to the water dissociation and adsorption at the {01$\bar{1}$N} and {01$\bar{2}$N} surfaces, which decreased the numbers of dangling bonds. Here, only low-index surfaces are studied on the consideration that low-index surfaces are much more stable than high-index surfaces and that there is no experimental evidence of generation of high-index surfaces during reactions (Fig. 7.18).

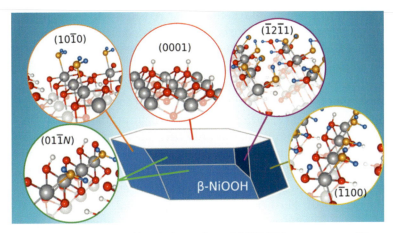

Figure 7.18 Represented low-index surface of β-NiOOH in aqueous condition.

Owing to its simplicity and great importance, the OER mechanism catalyzed by NiOOH has been well studied. Considering that the (0001) surface is the most stable, OER takes place on the (0001) surface most likely. Carter calculated the single-site association mechanism, binuclear H_2O–O mechanism, binuclear OH–OH mechanism, and binuclear H_2O_2 mechanism on the (0001) surface of β-NiOOH and found that the binuclear H_2O_2 mechanism is energetically highly unfavorable, whereas the binuclear H_2O–O mechanism and the binuclear OH–OH mechanism have the lowest overpotentials (~0.5 V), lower than the single-site association mechanism (~0.6 V). However, as mentioned in Section 7.2, the binuclear mechanism is kinetically unfeasible.

Nørskov and Bell performed DFT calculations to study the OER mechanism catalyzed by β-CoOOH, and it was found that the ($10\bar{1}4$) surface (η = 480 mV) gives a lower overpotential than the ($01\bar{1}2$) surface (η = 800 mV). The ($10\bar{1}4$) surface is abundant in Co^{3+} ions, whereas the ($10\bar{1}4$) surface has more Co^{4+} ions. Association of OH^- to generate OH* would oxidize Co^{2+} to Co^{3+} on the ($10\bar{1}4$) surface, whereas it oxidizes the Co^{3+} to Co^{4+} on the ($01\bar{1}2$) surface. Transformation of Co^{3+} to Co^{4+} leads to a too-weak OH* binding, and the formation of OH* becomes the rate-determining step. However, for the ($10\bar{1}4$) surface, the formation energy of OH* has an optimal value of 1.23 eV.

There is a debate on whether Ni or Fe is the active site in NiFe LDH, and most work supports that Fe is the active site. The debate has been well reviewed by Kundu; therefore, we will not review the debate on the active sites of NiFe LDH. However, recently, Goddard performed extensive calculations using grand canonical quantum mechanics to have a better understanding of the synergy between Fe and Ni in (Ni,Fe)-OOH, which sheds light on the debate. They suggested that both Fe and Ni are active sites; high-spin d^4 Fe(IV) can stabilize the active O radical intermediate, whereas the low-spin Ni(IV) catalyzes the subsequent O–O coupling. Therefore, it is the synergy between Fe and Ni that leads to the optimal OER activity of (Ni,Fe)OOH. Furthermore, overpotential and Tafel slope were calculated to be 420 mV (Fig. 7.19) and 23 mV/dec, in good agreement with experiments, 300–400 mV and 30 mV/dec, respectively.

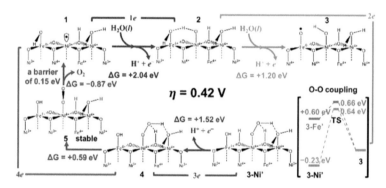

Figure 7.19 Calculated OER mechanism catalyzed by NiFe LDH by Goddard.

Having established the idea that Fe could stabilize the O radical thus facilitating OER, Goddard further performed DFT calculations to substitute Fe with other transition metals (groups 3–9). And it was found that Co-, Rh-, and Ir-doped NiOOH could better stabilize O radical and exhibited even lower overpotentials, 270, 150, and 20 mV, respectively. Although NiCo LDH was proposed to have a better OER activity than NiFe LDH by Goddard, it actually showed a poorer performance than NiFe LDH, which probably is due to the smaller stabilization ability of Co with the formed oxy (O*) group because of the larger number of electrons in the antibonding orbitals of metal and oxygen π bonds, as shown in Fig. 7.20. It can be expected that the lower number of electrons in the d-orbitals could lead to stronger metal-oxy bond; however, if the metal-oxy bond is too strong, it is also unfavorable for OER, and that could be the reason why NiMn LDH and NiTi LDH exhibited poor OER activities.

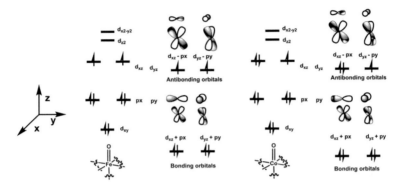

Figure 7.20 Molecular orbital analysis of Fe=O and Co=O containing complexes.

As Ni^{4+} is responsible for catalyzing O–O coupling due to its strong oxidizing ability, one can expect that if the oxidation ability of metal is decreased, the ability of metal to catalyze the O–O coupling should also decrease. In the first-row transition metal, oxidation ability of metal ions with the same positive charge would decrease because of its lower electronegativity. Indeed, changing Ni to Co, we can see a decrease in OER activity, and that is why CoFe LDH exhibited a lower OER activity than NiFe LDH.

7.6 Summary and Prospects

In summary, the nonmetal element doping effect has received considerable interest from the research community, which has triggered the exploration of the fascinating electrochemical properties of transition-metal compounds. There is not a doubt that this research wave will continue. In this chapter, we have taken stock of methods that have been developed for the synthesis of binary-nonmetal TMCs, including the conventional pyrolysis reaction, chemical vapor deposition, solid/gas-phase doping, facile hydrothermal/solvothermal methods, solid-state methods, and hot-injection methods.

It has been found that introducing foreign nonmetal atoms into transition-metal compound lattices can considerably tune the electronic structures of the parent TMCs and shift the hydrogen adsorption free energy toward the thermal neutral direction. Consequently, such efforts have resulted in superior catalytic performance for the HER. Meanwhile, incorporating bigger or smaller radii of foreign nonmetal atoms can cause a distortion of TMC lattices, resulting in favorable bond breaking of the adsorbed molecules, thus increasing active sites. Moreover, alongside the electronic structure modulation, the bandgap narrowing in binary-nonmetal TMC systems has also been successfully observed by the DFT calculations, resulting in enhanced conductivity compared to that of the parent TMCs, which can facilitate chemical-reaction kinetics and thereby boost the HER electrocatalysis.

Although extensive efforts have been made in the research of binary-nonmetal TMC materials for the HER, there still remain challenges to overcome in this exciting area, which we will discuss as follows.

7.6.1 Difficulty in Synthesizing Well-Defined Binary-Nonmetal TMC Materials

From the synthesis point of view, it is still quite difficult to synthesize well-defined binary-nonmetal TMC materials with uniform distribution of foreign nonmetal atoms and desired compositions. To obtain uniform and stoichiometric compositions, using a solid-state synthesis method seems to be a way, but it needs a high reaction temperature and long reaction time. Great efforts are still needed to develop an efficient way to synthesize binary-nonmetal TMC materials with uniform and desired compositions.

7.6.2 Difficulty in Characterizing Precise Locations of As-Doped Foreign Nonmetal Atoms in Binary-Nonmetal TMC Materials

Although various ways have been reported showing the successful incorporation of foreign nonmetal atoms, it is still very difficult to characterize the precise locations of the As-doped foreign nonmetal atoms, making it practically difficult to understand how foreign nonmetal atoms modulate the electronic structure and thus the physicochemical properties of TMC catalysts. In this regard, state-of-the-art spectral and microscopy characterization techniques, such as X-ray adsorption spectroscopy (XAS), HRTEM, STEM, and scanning tunneling microscopy (STM), as well as theoretical calculations may help to unveil the hidden connection between the foreign nonmetal doping and the modulated electronic structure.

7.6.3 Need for Further Improvement in Their HER Electrocatalytic Performance

Although binary-nonmetal TMC catalysts, which hold great promise as low-cost alternatives to noble metal catalysts, have achieved better HER catalytic performance than their parent TMC catalysts, most current-application-qualified HER catalysts are still based on noble metals such as Pt. Therefore, further improvement in their HER electrocatalytic performance is clearly needed. For practical applications, desirable electrochemical catalysts should have a low overpotential and low Tafel slope, which can be used as a descriptor to guide the design and fabrication of better noble-metal-free HER electrocatalysts. The HER performance characteristics, including the overpotential and Tafel slope of some typical binary-nonmetal TMC catalysts, such as foreign nonmetal doped transition-metal phosphides (X-MP), foreign nonmetal doped transition-metal nitrides (X-MN), and foreign nonmetal doped transition-metal sulfides (X-MS), are summarized and compared in Fig. 7.10. One can see that compared to the foreign nonmetal doped transition-metal nitrides and sulfides (X-MN and X-MS), the foreign nonmetal doped transition-metal phosphides (X-MP) exhibit both lower overpotentials and lower Tafel slopes. Future studies should be directed toward the elucidation of the mechanisms underpinning these results. Therefore, more careful studies should emphasize the interplay between theoretical calculations and experimental investigations to address important issues on structural design and rational fabrication with precise control over the morphology and crystal structure of HER catalysts. The informed design should take into account issues such as interfacial adsorption and atomic/

molecular transportation in the hydrogen evolution process. These efforts are expected to further improve the HER electrocatalytic performance of X-MP and to realize their promise as alternatives to Pt-based HER catalysts.

We also summarized the recent developments on OER catalyzed by LDH. The basic criteria to evaluate the performance of OER catalyst have been presented. Unary metal-based LDH has been discussed to have a deep understanding of the intrinsic activity of LDH. Basically, unary metal-based LDH is not effective for OER due to its poor conductivity and absence of synergistic effect. On the other hand, binary metal-based LDH is much more OER active. In particular, NiFe LDH is the most effective binary metal-based LDH toward OER. It outperformed other Ni-based binary metal LDHs owing to its suitable M–OH bond strength, which is neither too strong nor too weak. And it exhibited better OER activity than non-Ni-based LDH because Ni^{4+} has a strong oxidation ability, which can facilitate the formation of O–O bond. Finally, ternary metal-based LDH shows better OER activity owing to its increased conductivity. Moreover, introducing divalent defects leads to exposure of the Ni-O-Fe unit, which would greatly increase the OER activity of LDH.

Although LDH as a promising OER catalyst has been well studied over the past few years, some critical issues should be tackled to realize the practical application of these catalysts on hydrogen production via electrochemical water splitting.

1. It is known that LDH with a high OER activity almost always has a very high conductivity, but the underlying mechanisms are still not understood. Moreover, it has been noted that the conductivity of LDH depends on the applied potential. Therefore, the influence of conductivity on the OER activity of LDH catalysts should be addressed systematically, especially the intrinsic conductivity of LDH, the contact resistivity between LDH and substrate, and the applied potential-dependent conductivity.

2. The synergistic effect between transition metals should be studied further to obtain the fundamental details, instead of simply going by the observation that the LDHs with more different transition metals are more OER active. Nowadays,

the synergistic effect between Ni and Fe in NiFe LDH seems to be clearer, but the synergistic effect between other binary transition-metal-based LDHs and ternary transition-metal-based LDHs remains to be elucidated.
3. Close interplay between theoretical and experimental studies should be encouraged to tackle not only the thermodynamics but also the kinetics and dynamics of the intermediate steps of OER catalysis. For example, the key transition states leading to the formation of OOH* and O* should be located to have a better understanding of the kinetics of OER catalysis. For most LDHs, transition metals will be peroxided before OER takes place. Thus, pre-oxidations should be taken into considerations in future studies to have a better understanding of the active phase of LDH, which is usually ignored in the current practice.
4. Most OER experiments are performed in highly basic condition (pH > 13), but the OER performed in acidic and neutral conditions with more practical utilization is less reported and thus should be studied further in the future.

Questions

1. Compare the advantages and disadvantages of different water hydrogen production methods and put forward their own views on their development prospects.
2. What are the main types of catalysts for hydrogen production by water electrolysis? What are the main preparation methods and their advantages and disadvantages?
3. What needs to be improved for membrane electrodes used for hydrogen production from electrolytic water?
4. If you are a businessman seeking to invest in hydrogen production technology, which hydrogen production method will you choose and what is the reason?
5. What are the mechanisms of OER? What are the basic criteria for choosing OER catalysts?
6. What are the advantages and disadvantages of the electrocatalysts based on transition-metal-based layered double hydroxides for oxygen evolution?

References

Elbert K., Hu J., Ma Z., Zhang Y., Chen G., An W., Liu P., Isaacs H. S., Adzic R. R., and Wang J. X. (2015). *ACS Catal.*, **5**, 6764–6772.

Hu J., Zhang C. X., Meng X. Y., Lin H., Hu C., Long X., and Yang S. Y. (2017). *J. Mater. Chem. A*, **5**(13), 5995–6012.

Li H., Tsai C., Koh A. L., Cai L. L., Contryman A. W., Fragapane A. H., Zhao J. H., Han H. S., Manoharan H. C., Abild-Pedersen F., Norskov J. K., and Zheng X. L. (2016). *Nat. Mater.*, **15**, 48–54.

Long X., Wang Z. L., Xiao S., An Y. M., and Yang S. (2016). *Mater. Today*, **19**(4), 213–226.

Morales-Guio C. G., Stern L. A., and Hu X. L. (2014). *Chem. Soc. Rev.*, **43**, 6555–6569.

Voiry D., Salehi M., Silva R., Fujita T., Chen M. W., Asefa T., Shenoy V. B., Eda G., and Chhowalla M. (2013). *Nano Lett.*, **13**, 6222–6227.

Wang Z., Long X., and Yang S. H. (2018). *ACS Omega*, **3**(12), 16529–16541.

Zheng Y., Jiao Y., Jaroniec M., and Qiao S. Z. (2015). *Angew. Chem. Int. Ed.*, **54**, 52–65.

Index

active layer 52
anode materials 118, 119, 122–124

Bi_2O_3 26–29, 46, 48
Bi-based perovskites 179
binary metal-based 226, 233, 242
bottom-up 10–12, 36, 37

carbon dots 9–11, 48, 111, 116
carbon quantum dot 53, 91–116
carbon-based electrode 82
carbonitrides 215, 217
catalyst 12, 13, 36, 190–242
catalytic materials 5, 190
cathode materials 130, 138, 141
charge dynamics 50, 52
clean energy 1, 7, 9, 37, 46–49, 91, 189
copper hydroxide 18–21
copper sulfide 12, 13, 15, 46
coprecipitation 36, 43, 119, 231, 233
C-PSCs 78–87, 158, 159, 173, 174
CQDs 92, 99–116
$CsGeI_3$ 174, 175
$CsPbBr_3$ 42–44, 145, 148, 150, 154, 155, 159
$CsPbI_3$ 145–154, 159–165, 185
$CsPbI_{3-x}Br_x$ 159–163
$CsSnBr_3$ 148, 169–172, 177
$CsSnI_3$ 145–148, 165–177, 185
$CsSnI_{3-x}Br_x$ 172
Cu_2S/Au core/sheath 23–25
Cu_2S/polypyrrole 25, 46
current–voltage hysteresis 67

direct bandgap 90, 95, 96, 98, 181, 186
domains 38, 101, 103, 205
double perovskites 182
dye-sensitized solar cells 12, 55

electrocatalytic water splitting 7, 189
energy band 5, 60, 66, 92
energy conversion 1, 3–6, 9, 47, 113, 118, 147, 148, 189
energy-level alignment 50–52, 70, 71, 77, 81
ETL 49, 53, 57–69, 75, 78–80, 88
extrinsic 93, 94

fluorescence 10, 11, 45, 99–116

grain boundaries 59–65, 68, 73, 75, 76, 163

HER 5, 37, 47, 190–242
hierarchically 117–143
hot-injection 40–43
HTL 49, 58, 65–73, 78, 88
hydrogen adsorption 190, 193–214, 240
hydrogen evolution 37, 47, 190, 191, 199, 202–217, 242

indirect bandgap 92, 95, 96, 99, 116, 182, 186
inorganic trihalide perovskite 9, 40–47
interface engineering 6, 7, 49–89

interfaces 1–7, 49–60, 67, 68, 79, 87–91, 117, 145, 146, 172, 187, 189
intrinsic 37, 53, 93, 100, 101, 106, 111, 209, 212, 215, 221, 224–228, 242
ion migration 53, 54, 65, 68, 75, 79, 80, 88, 89, 159, 164
I-PVKs 145–149, 159, 181–186
iron oxide 17, 130, 134

layered double hydroxides 9, 34, 35, 47, 111, 191, 192, 223–233, 243
LDH 5, 34–37, 192, 221–243
lithium-ion battery 117–142
luminescent materials 7, 9, 91–116

materials and interfaces 1–9, 49, 79, 91, 117, 145, 146, 189
metal carbonate 120–123, 127, 129, 140
metal carbonate hydroxide 120, 127, 129
metal hydroxide 35, 119, 120, 127, 129, 140
metal–organic framework 130
microfluidic 12, 41, 45

nanomaterials 3, 7, 9, 12, 92, 101, 110, 118, 121, 134, 141, 214, 217
nanostructures 12, 13, 18, 22, 32, 48, 56, 117–129, 134, 140–143, 196
nanotubes 3, 10, 12, 22, 25, 32–34, 39, 47, 85, 119, 206, 227
n-i-p 49, 55–68, 78, 88, 169

OER 5, 191, 192, 209, 214–243
oxygen evolution 191, 215, 223–233, 243

oxyhydroxides 224, 226

perovskite solar cells 7, 49, 57, 87, 89, 145–187
phosphorescence 100, 108–113, 116
phosphosulfides 209, 210, 212, 214
photoactive materials 3, 4, 36
p-i-n 49–53, 61, 64–89
p-n junction 92–95, 116
precursors 10, 11, 35–38, 44–47, 118–142, 150, 162, 208, 235
PSCs 49–88, 145–186

reaction pathways 198–201
recrystallization 41, 44, 45, 129
reprecipitation 41, 43
room temperature phosphorescence 100, 108, 109, 111, 116
RTP 100, 110–113, 115

Sb-based perovskites 181
semiconductors 5, 9, 40, 50, 69, 92–98, 116, 146, 212
silver sulfide 16
Sn-based perovskites 175
solvothermal 11, 40, 41, 46, 113, 125, 127, 227, 239
sulfoselenides 202
surface defect-derived 106

TCO 49, 57, 58, 68
ternary metal-based 233, 242
tetrapods 30–32, 47, 48
thermally activated delayed fluorescence 100, 113
TADF 100, 111, 109, 123–125
TMCs 9, 37–40, 47, 190–192, 201–217, 239, 240

top-down 10, 36, 37, 46, 47
transition-metal-based 130, 191, 223, 225–243
trap passivation 50, 54

unary metal-based 223, 233, 242

up-conversion 99, 107, 116

valence electron 1, 3, 4, 7, 210, 212
V-hydroxides 226

ZnO 17, 22, 23, 30–34, 46–49, 55, 56, 67, 78, 127–129, 176, 177